How to Ace the Rest of Calculus:
The Streetwise Guide

How to Ace
the Rest of Calculus:
The Streetwise Guide

Colin Adams
Williams College

Joel Hass
University of California, Davis

Abigail Thompson
University of California, Davis

W. H. Freeman and Company
New York

Cover Designer: Cambraia F. Fernandes
Text Designer: Victoria Tomaselli

Library of Congress Cataloging-in-Publication Data

Adams, Colin Conrad.
  How to ace the rest of calculus : the streetwise guide / Colin Adams, Joel
Hass, Abigail Thompson.
  p. cm.
  Includes index.
  ISBN 0-7167-4174-1
  1. Calculus–Study and teaching. I. Hass, Joel. II. Thompson,
Abigail. III. Title.
  QA303.3.A332 2001
  515–dc21   00-066292
                                                                    CIP

Printed in the United States of America

First printing 2001

W. H. Freeman and Company
41 Madison Avenue, New York, NY 10010
Houndmills, Basingstoke RG21 6XS, England

**To all those students who
bought our first book.
We sure hope you got an A.**

CONTENTS

Introduction

This is the second book in the series *How to Ace _____ : The Streetwise Guide*. This volume is for students who have already taken at least one semester of calculus. It contains everything you are likely to need to know about the rest of calculus, including, among other things, multivariable calculus and sequences and series. Why did we write it? Because of the mountain of letters of support for the first book in the series, *How to Ace Calculus: The Streetwise Guide,* as illustrated by the following unsolicited letter:

Dear Math Types,

Who would have thought that three pencil-necked number heads could come up with a book that I would enjoy reading? But there you have it. Your *How to Ace Calculus: The Streetwise Guide* was a great read. In fact, I find that I like mathematics so much now that in order to give myself more time for it, I am going to give up being chair of the Hazing Committee at my fraternity. I am even considering giving up beer. Of course, then I would have to resign from the frat. But hey, I don't think I would really care anymore. Having had my eyes opened to the sublime beauty of mathematics, I have a newfound respect for the creative endeavors of humanity. This brings with it a respect for each and every individual in our diverse society. I no longer find myself judging people based on their outward appearance, such as the fact that their necks resemble pencils. I realize that physical characteristics do not reflect the immense creative potential that resides in every human being. So I thank you for changing my life. I will now dedicate myself to eradicating just the kind of attitude that I myself exemplified. I can only hope that in my small way, I can make a contribution to society that even approaches the tremendous contribution you three have made through your book. You are truly exceptional people, and I will be forever in your debt. Please feel free to call on me at any time if I can help you in any way, personally, financially, or otherwise.

With the warmest of regards,

Billy Bob Wainger

P.S.: I can only hope and pray that you will consider writing a sequel to that first book that covers the rest of calculus, including sequences and series and multivariable calculus.

Okay, okay, so it isn't a real letter. We made it up. But what is important is the sentiment it expresses, a sentiment that was expressed in bag after bag of mail that we received regarding our first book. And we are thrilled that so many people expressed such strong feelings about our necks.

If you are looking at this book in the bookstore, then either:

1. You took beginning calculus and used our previous book and got an A. What are you waiting for? It worked once, didn't it?

2. You took beginning calculus and didn't use our previous book and got a poor grade. Clearly you should get this book to rectify your previous mistake. Can't hurt to buy the first book too.

3. You didn't buy our previous book and still got an A. Well, did you ever stop to think that maybe you got lucky? Better buy this book to stack the odds in your favor.

4. You bought the first book and didn't get an A. Did we forget to mention you have to go to class and take the exams?

We would hate to think you're standing around at the bookstore reading this, in a crowded aisle with nowhere to sit down and relax. You would probably be a lot more comfortable at home, sipping a cold soft drink as you casually peruse your new purchase.

We're not going to repeat all the advice we gave in the first book on how to pick your instructor, how to study, and all of that nonsense. (Although we hope that book is on your bookshelf next to this one.) The same rules apply.

This book is devoted to the specific topics of the rest of calculus. It won't replace your text. It can't. It doesn't weigh enough. But it will explain what is really going on in the course. So go to it, and have fun!

P.S.: Check out our Web site **howtoace.com** where you will find lots of stuff to help you get that A, including links to exam problems, math jokes, and additional explanations.

Indeterminate Forms and Improper Integrals

2.1 Indeterminate forms

Knowing your limits is a very important skill. You have no business hanging by your fingertips from the roof of the World Trade Center unless you know you can do so safely. You have to have put in a lot of practice beforehand, hanging from the jungle gym during recess, then hanging from the bathroom windowsill, then the hometown watertower, the Space Needle in Seattle, and finally the World Trade Center. You must know what you are capable of, what your limits are.

Similarly, in math, you want to be able to figure out your limits. Given $\lim_{x \to 3} x^2 + 2$, you want to know that it equals 11. Or that

$$\lim_{x \to e} \frac{\ln x - 2}{x^2} = \frac{-1}{e^2}$$

But just as you may overestimate your finger strength and find your situation dire, so may you overestimate your ability to determine limits. For instance, what is $\lim_{x \to 2} \frac{3x^2 - 12}{x^3 - 8}$?

Not so obvious, is it? If we plug in $x = 2$, we get $\frac{0}{0}$. Is $\frac{0}{0}$ equal to $0, 1, \infty$? Good question. In fact, the limit could be any of these. That's why we call a limit like this an *indeterminate form.*

The most notorious indeterminate form is

$$\lim_{x \to 0} \frac{\sin x}{x}$$

If we plug in $x = 0$, we get $\frac{0}{0}$. But we really need to know this limit. It comes up in a variety of situations, in particular when trying to show that the derivative of $\sin x$ is $\cos x$ using the limit definition of the derivative. There is a rule that can be used to find such limits. This rule was discovered by Johann Bernoulli, one of the great mathematicians of the seventeenth century, but it is named for the Marquis of L'Hôpital, who was paying Bernoulli to teach him calculus and taking credit for Bernoulli's results. In fact, we authors are doing that with this book. We haven't written a word of this, although our names are on the cover. We're paying Bill Gates to write the whole thing. L'Hôpital is pronounced Low-pee-tall.

Anyway, here's

L'Hôpital's rule   if $\lim\limits_{x \to a} \dfrac{f(x)}{g(x)}$ is an indeterminate form, then

$$\lim_{x \to a} \frac{f(x)}{g(x)} = \lim_{x \to a} \frac{f'(x)}{g'(x)}$$

What is this rule trying to tell us? It says, "If you don't know what happens to $\dfrac{f(x)}{g(x)}$ when you take the limit, just look at $\dfrac{f'(x)}{g'(x)}$ instead!"

Example 1   Find $\lim\limits_{x \to 0} \dfrac{\sin x}{x}$.

Solution   First we check it's an indeterminate form by plugging $x = 0$ into the expression $\dfrac{\sin x}{x}$ and getting $\dfrac{0}{0}$. So L'Hôpital's rule applies:

$$\lim_{x \to 0} \frac{\sin x}{x} = \lim_{x \to 0} \frac{(\sin x)'}{x'} = \lim_{x \to 0} \frac{\cos x}{1} = \frac{1}{1} = 1$$

Wow, was that easy! You have to love this rule. Let's try another!

Example 2   Find $\lim\limits_{x \to 1} \dfrac{x^2 - 1}{x - 1}$.

Solution   This is an indeterminate form $\dfrac{0}{0}$. So we can apply L'Hôpital to obtain:

$$\lim_{x \to 1} \frac{x^2 - 1}{x - 1} = \lim_{x \to 1} \frac{(x^2 - 1)'}{(x - 1)'} = \lim_{x \to 1} \frac{2x}{1} = 2$$

In fact, we could actually do this without L'Hôpital. We could just do a little algebra to clean it up, like this:

$$\lim_{x \to 1} \frac{x^2 - 1}{x - 1} = \lim_{x \to 1} \frac{(x - 1)(x + 1)}{x - 1} = \lim_{x \to 1} (x + 1) = 2$$

Hey, whatever steams your clams.

Example 3   Find $\lim\limits_{x \to 0} \dfrac{e^x - x - 1}{x^2}$.

Solution   Again an indeterminate form $\dfrac{0}{0}$! They're everywhere. L'Hôpital says

$$\lim_{x \to 0} \frac{e^x - x - 1}{x^2} = \lim_{x \to 0} \frac{e^x - 1}{2x} \qquad \text{(Still of form } \frac{0}{0}\text{, so L'Hôpital's}$$
$$= \lim_{x \to 0} \frac{e^x}{2} = \frac{1}{2} \qquad\qquad \text{rule applies again.)}$$

Notice we applied L'Hôpital's rule twice there. But we had to check that we still had an indeterminate form before we applied it the second time.

Common Mistake   Applying L'Hôpital's rule to a form which is not indeterminate is a common mistake. This error can occur in any situation, but it's especially common when L'Hôpital's rule is being applied multiple times in a problem. Be sure to check at each stage that it still applies.

INDETERMINATE FORMS INVOLVING ∞

L'Hôpital's rule can also be applied if we have the indeterminate form $\dfrac{\infty}{\infty}$.

Example 4   Find $\lim\limits_{x \to \infty} \dfrac{\ln x}{\sqrt{x}}$.

Solution If we evaluate the numerator and denominator at $x = \infty$, we get $\frac{\infty}{\infty}$. So this is an indeterminate form. Let's apply L'Hôpital:

$$\lim_{x \to \infty} \frac{\ln x}{\sqrt{x}} = \lim_{x \to \infty} \frac{1/x}{1/(2\sqrt{x})} = \lim_{x \to \infty} \frac{2}{\sqrt{x}} = 0$$

The limit is 0. How do we interpret that? This just means that the denominator $\sqrt{x}$ grows faster than the numerator $\ln x$ as x goes to ∞.

OTHER INDETERMINATE FORMS

We have seen that if a limit is of the form $\frac{0}{0}$ or $\frac{\infty}{\infty}$, then we can apply L'Hôpital's rule. But that is the only time it applies.

Warning $\frac{0}{\infty}$ is not an indeterminate form. It is 0. And $\frac{\infty}{0}$ is not an indeterminate form. It is $\pm\infty$. In neither case does L'Hôpital's rule apply.

Sometimes other forms can be manipulated into the form $\frac{0}{0}$ or $\frac{\infty}{\infty}$. For example, suppose $\lim_{x \to a} f(x) \cdot g(x) = 0 \cdot \infty$. This is also an indeterminate form, since this product could be anything. But it's easy enough to turn this into $\frac{0}{0}$ or $\frac{\infty}{\infty}$ by writing:

$$f(x) \cdot g(x) = \frac{f(x)}{1/g(x)} \qquad \text{or} \qquad f(x) \cdot g(x) = \frac{g(x)}{1/f(x)}$$

Then it's straightforward to solve.

Example 5 Find $\lim_{x \to \infty} e^{-x} \ln x$.

Solution Letting x go to ∞, we see that this is of the form $0 \cdot \infty$. So we rewrite it as

$$\lim_{x \to \infty} e^{-x} \ln x = \lim_{x \to \infty} \frac{\ln x}{e^x}$$

This is now the indeterminate form $\frac{\infty}{\infty}$, so

$$\lim_{x \to \infty} \frac{\ln x}{e^x} = \lim_{x \to \infty} \frac{1/x}{e^x} = 0$$

Hey, whatever burps your baby.

2.2 Improper integrals

Well, we tried to avoid it, but now we have to introduce them. You know the type. They wear a T-shirt with a tux printed on it to the Calculus Cotillion. They eat the entire dinner with the dessert fork. They laugh loudly at involuntary auditory signs of digestive distress. That's right, we're talking about the improper integrals. If they weren't so important, they wouldn't get invited at all.

There are a couple of variations on this theme. First, there is the integral with unbounded interval of integration. That's right. Instead of integrating with limits from -2 to 4 or 3 to 7, these integrals have limits from 1 to ∞ or $-\infty$ to 2, or even $-\infty$ to ∞. For example, we could have the integral $\int_1^\infty \frac{1}{x^2}\, dx$.

A second type of improper integral is one where the function being integrated goes to $\pm\infty$ somewhere over the interval of integration. Take a look at $\int_{-1}^2 \frac{1}{x^{2/3}}\, dx$, which has this problem at $x = 0$.

But let's not swallow the entire fish whole. We'll start with the first type.

INFINITE LIMITS OF INTEGRATION

Let's look at the example $\int_1^\infty \frac{1}{x^2}\, dx$. It seems a little strange having a limit of integration that is infinite. What are we to make of it? Well, we know by now that definite integrals often represent areas under graphs of functions. So let's go with that viewpoint and see what it says in this case. It still makes sense to talk about the area under the curve $1/x^2$ for x going from 1 to ∞ (see Figure 2.1).

Now we know what you're thinking. That area goes on forever as we head out along the x-axis, so this integral must give ∞ for the answer. Easy! But hold on to your brain pan, because in fact, that's not what happens! To see what really happens, we need to figure out the correct interpretation of that upper limit of integration ∞. The right way to think of ∞ here is as a limit of

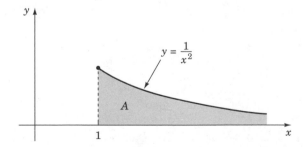

Figure 2.1 The area corresponding to $\int_1^\infty \frac{1}{x^2}\, dx$.

numbers which are getting very large, or $\lim\limits_{b \to \infty} b$. So the official interpretation of our improper integral is

$$\int_1^\infty \frac{1}{x^2}\,dx = \lim_{b \to \infty}\int_1^b \frac{1}{x^2}\,dx = \lim_{b \to \infty} \frac{-1}{x}\Big|_1^b = \lim_{b \to \infty}\left(\left(\frac{-1}{b}\right) - (-1)\right) = 0 - (-1) = 1$$

So in this case, that area actually turned out to be 1.

We will always interpret an infinite limit improper integral this way:

$$\boxed{\int_a^\infty f(x)\,dx = \lim_{b \to \infty}\int_a^b f(x)\,dx}$$

Sometimes an improper integral gives a finite number, as happened above. Then we say the improper integral *converges*. But sometimes the limit is ∞ or doesn't exist. Then we say the improper integral *diverges*.

Example   Find $\displaystyle\int_1^\infty \frac{1}{x}\,dx$.

Solution   This doesn't look very different from the previous example, but whammo, when we take a limit to compute it, we get

$$\int_1^\infty \frac{1}{x}\,dx = \lim_{b \to \infty} (\ln x)\Big|_1^b = \lim_{b \to \infty}((\ln b) - (0)) = \infty$$

This one *diverges*.

We also define

$$\boxed{\int_{-\infty}^b f(x)\,dx = \lim_{a \to -\infty}\int_a^b f(x)\,dx}$$

and

$$\boxed{\int_{-\infty}^\infty f(x)\,dx = \int_{-\infty}^c f(x)\,dx + \int_c^\infty f(x)\,dx}$$

for any real number c that we want to use. We can use a birthdate or a lucky number, or 0. We'll always get the same answer, so we can use whatever is easiest to compute with.

Now for the second type of improper integral.

INFINITE INTEGRANDS

Given the chore of computing the integral

$$\int_{-1}^{2} \frac{1}{x^{2/3}}\, dx$$

we certainly have a problem, since $\frac{1}{x^{2/3}}$ is undefined at $x = 0$. It explodes right there, one big volcano (see Figure 2.2).

On the other hand, the area under the curve doesn't look that big. High yes, but not that wide.

To make sense of this, we split the integral into two integrals around the problem point.

$$\int_{-1}^{2} \frac{1}{x^{2/3}}\, dx = \int_{-1}^{0} \frac{1}{x^{2/3}}\, dx + \int_{0}^{2} \frac{1}{x^{2/3}}\, dx$$

Now, we will use the same idea as above. Since it doesn't make sense to talk about $1/x^{2/3}$ at $x = 0$, we replace the 0 in the first integral by a limit.

$$\int_{-1}^{0} \frac{1}{x^{2/3}}\, dx = \lim_{b \to 0^{-}} \int_{-1}^{b} \frac{1}{x^{2/3}}\, dx$$

We take the limit as b approaches 0 from the left, since the interval of integration is all to the left of 0. Then we can compute the integral and take

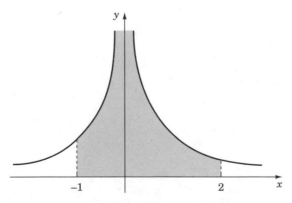

Figure 2.2 The area given by $\int_{-1}^{2} \frac{1}{x^{2/3}}\, dx$.

the limit:

$$\lim_{b \to 0^-} \int_{-1}^{b} \frac{1}{x^{2/3}} \, dx = \lim_{b \to 0^-} 3x^{1/3} \Big|_{-1}^{b} = \lim_{b \to 0^-} [3b^{1/3}] - [3(-1)^{1/3}] = 3$$

Similarly,

$$\lim_{b \to 0^+} \int_{b}^{2} \frac{1}{x^{2/3}} \, dx = \lim_{b \to 0^+} 3x^{1/3} \Big|_{b}^{2} = \lim_{b \to 0^+} [3(2^{1/3})] - [3b^{1/3}] = 3(2)^{1/3}$$

So

$$\int_{-1}^{2} \frac{1}{x^{2/3}} \, dx = 3 + 3(2)^{1/3} \approx 6.780$$

Hey, whatever roasts your chestnuts.

Polar Coordinates

3.1 Introduction to polar coordinates

You know how in all those Cold War movies where we're submerged in a submarine and there's a Soviet sub just 50 feet away, so everybody has to be really quiet or else we'll get a torpedo right up the torpedo tube? And there's this one ensign sitting in front of the the radarscope, and on the scope there is this green radial ray that goes round and round. And all of the sailors are sweating like they just took showers in their clothes, because after all, they are packed into a sardine can and there's not enough room to bring inessentials like deodorant. And the captain whispers, "Not a sound. Nobody make a sound." And all the while, that line on the radarscope is going round and round. And each time it goes round, a big green dot corresponding to the enemy sub shows up on the screen. And every time that line goes round the scope beeps. Yes, that's right. It beeps like an alarm clock going off in the middle of the night. And you just have to wonder, sitting there in front of the TV, why can't the Russians hear the stupid scope? Do they have cotton stuffed in their ears? Are they confusing the beep of that scope with their own? The fact of the matter is that scope is loud enough to wake the dead. And as you sit in your home theater, you want to say to the captain, "You don't have to whisper. Nobody could hear you over the damn scope." You could sing God Bless America with your tonsils popping out the back of your throat, and the Russians would say, "Comrade, did you hear something?" "Comrade, I can't hear anything over that damn radarscope."

And of course, in at least 80 percent of these movies, the action is taking place under the polar ice cap. This is where submarines tend to bump into one another. So it's no big surprise that the coordinates used for the scope are called polar coordinates.

The ensign whispers to the captain in a very loud whisper, in order to be heard over the scope, "Captain, looks like a C class nuclear ranger. Appears to be carrying 37 men, 12 women, and one free-range chicken. She's 50 feet away and closing."

The scope always puts your own position at the center. Your first measurement of the other point is its distance from you, in this case 50 feet.

Then Radar Boy adds, "She's at 37 degrees," meaning that she is 50 feet out, at an angle of 37 degrees from the positive x axis. We would say that the polar coordinates of the point are $(r, \theta) = (50, 37°)$.

Of course, unlike the navy, we use radians, because all mathematicians have agreed always to use radians. It makes our computations easier. If all you use angles for is to aim an occasional torpedo, then degrees are fine. But if you want to do hard-core mathematics with the big dogs, then radians are the way to go.

To describe a point in the plane in polar coordinates, we give its distance from the origin, called r, and its angle in radians going counterclockwise from the positive x-axis, which we call θ. So, instead of using (x, y), we describe a point by (r, θ), as in Figure 3.1.

In polar coordinates, a point can have more than one description. The point $(1, 0)$ in rectangular coordinates has distance 1 from the origin and has

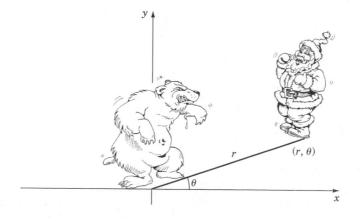

Figure 3.1 Polar coordinates determine Santa's position.

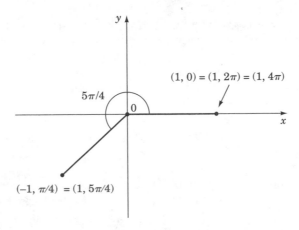

Figure 3.2 A single point is described in polar coordinates in more than one way.

angle 0 with the positive x-axis. So in polar coordinates, it is $(r, \theta) = (1, 0)$. But it is also given by $(1, 2\pi)$, $(1, 4\pi)$, etc. Even wackier is the origin, which is given by $(0, \pi)$, $(0, \pi/2)$, $(0, 7.3\pi)$, etc.

If r is negative, we go out from the origin in the opposite direction. So for instance, in Figure 3.2, $(r, \theta) = (-1, \pi/4)$ is the same as the point given by $(1, 5\pi/4)$.

From Figure 3.3 and the Pythagorean theorem, we see the following basic relationship between rectangular and polar coordinates:

$$x = r \cos \theta$$
$$y = r \sin \theta$$
$$r = \sqrt{x^2 + y^2}$$

Also, $\tan \theta = y/x$, assuming $x \neq 0$.

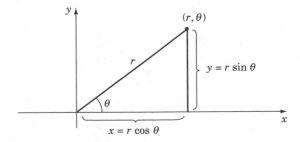

Figure 3.3 Rectangular coordinates in terms of polar coordinates.

These four equations are worth knowing, but you needn't memorize them, since they can so easily be read off the picture. Let's try graphing some polar equations.

Example 1 Graph $r = 3$.

Solution You're going to like this one. We want to draw all the points (r, θ) in the plane that satisfy $r = 3$. Since θ doesn't occur in the equation, θ can be anything at all. It's free as a bird, no constraints, no obligations. Anything goes, let 'er rip. On the other hand, r is more restricted, being forced as it is to be 3 at all times. But if r is fixed at 3, and θ can do what it wants, our ray of length 3 out of the origin can swing all the way around the full 360°. We get all the points on a circle of radius 3 around the origin. The graph appears in Figure 3.4.

The polar equation of this circle is simpler than the usual circle equation, $x^2 + y^2 = 9$. That's one reason why people are fond of polar equations. They make it incredibly easy to describe certain common graphs, particularly circles centered at the origin.

Example 2 Graph $r = 2 \sin \theta$.

Solution
Method 1 Plug in points. This is a time-honored tradition in graphing. When in doubt, plug in points. Not sure what's going on? Confused by the fancy pants techniques everybody's trying to cram down your throat? Then revert to your childhood, and plug in points, one at a time.

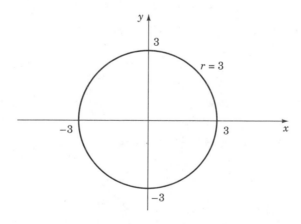

Figure 3.4 Graph of $r = 3$.

θ	0	$\pi/4$	$\pi/2$	$3\pi/4$	π	etc.
$r = 2\sin\theta$	0	$\sqrt{2}$	2	$\sqrt{2}$	0	etc.

This gives us Figure 3.5. Notice that we actually travel around the circle twice as we let θ vary from 0 to 2π.

Method 2   Bored with plugging in points? Looking for something that will exercise your big gray organ? The one in your head? How about this alternative. Multiply both sides of the equation by r to get $r^2 = 2r\sin\theta$. Then, since $r^2 = x^2 + y^2$ and $y = r\sin\theta$, we can convert the equation to rectangular coordinates, namely, $x^2 + y^2 = 2y$. Now what's this? Well, we can rewrite it as

$$x^2 + (y^2 - 2y + 1) = 1$$
$$x^2 + (y - 1)^2 = 1$$

This is a circle of radius 1 centered at (0, 1), as in Figure 3.5.

From this, it is easy to see that

The equation of a circle of radius a centered at $(0, a)$ is
$$r = 2a\sin\theta$$

Similarly,

The equation of a circle of radius a centered at $(a, 0)$ is
$$r = 2a\cos\theta$$

Examples of these circles appear in Figure 3.6.

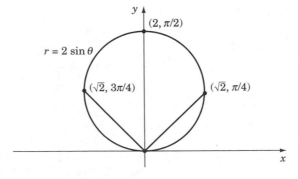

Figure 3.5 Graph of $r = 2\sin\theta$.

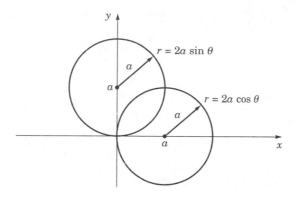

Figure 3.6 Circles given by $r = 2a \cos \theta$ and $r = 2a \sin \theta$.

Example 3 Graph $r = 1 + \sin \theta$.

Solution This one is trickier. If we try to convert to rectangular coordinates, we get a mess. We do better plugging in points on this one.

θ	0	$\pi/4$	$\pi/2$	$3\pi/4$	π	$5\pi/4$	$3\pi/2$	$7\pi/4$	2π
$r = 1 + \sin \theta$	1	1.707	2	1.707	1	0.293	0	0.293	1

The result appears in Figure 3.7. It is what is called a *cardioid,* since it resembles a heart. Not one of those stylized hearts from Valentine's Day that has only the slightest resemblance to the thing beating in your chest, but an actual version of the organ itself. You can tell it's a real heart because it has a Latin name.

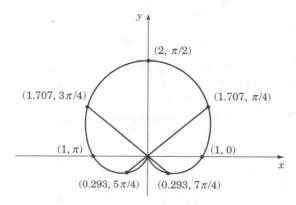

Figure 3.7 Graph of the cardioid $r = 1 + \sin \theta$.

Cardioids will be the result of any of the equations $r = a(1 \pm \sin \theta)$ or $r = a(1 \pm \cos \theta)$. When people say, "She has a heart as big as the great outdoors," they mean a is large. The choice of sin or cos determines if the heart is sideways or upright. Whether it's + or − determines whether the heart faces right side up, or upside down (or right side right or left).

These last few examples are the Who's Who of polar equations. They come up again and again, so it's probably worth learning them cold.

3.2 Area in polar coordinates

Just as curves in rectangular coordinates bound areas, so do curves in polar coordinates. Since we know how to find the area between curves given in rectangular coordinates, we could take curves described in polar coordinates, convert them to rectangular coordinates, and then find the areas. But this is akin to translating a poem by Robert Frost into Russian to decide how much we like it. Instead, we will see how to compute areas directly in polar coordinates.

Suppose that we want to find the area bounded by two radial lines given by $\theta = \alpha$ and $\theta = \beta$, and by the curve $r = f(\theta)$, for $\alpha \leq \theta \leq \beta$. (See Figure 3.8.)

This looks like a job for Riemann Sums, or put another way, time to "divide and conquer." We approximate the region by thin pie segments.

Speaking of pie, lemon meringue is a favorite dessert at math dinners. And when pie gets served, the first thing everybody does is discuss how to find the area of a piece of pie. (Actually, the first thing everyone does is check to see who got the biggest piece.) If the piece of pie has radius r and angle ϕ, as in Figure 3.9, then the whole pie has area πr^2 and the fraction of that

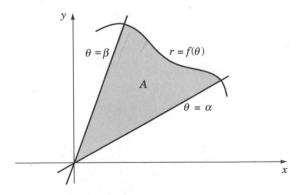

Figure 3.8 Area bounded by $\theta = \alpha$, $\theta = \beta$, and $r = f(\theta)$.

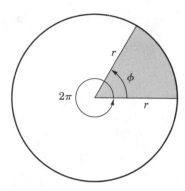

Figure 3.9 Finding the area of a piece of pie.

area in our particular piece is the fraction of the total angle 2π given by ϕ. That is to say, the area of the wedge is $\pi r^2 \times \dfrac{\phi}{2\pi} = \dfrac{1}{2}(r)^2 \phi$.

In case you were wondering what math dinners are like, now you know. You should see what happens when it comes to splitting the bill. Let's return to the problem at hand, finding the area of the region.

Take a look at Figure 3.10. We cut the angle between α and β into n smaller angles, each of size $\Delta\theta$, and we choose an angle θ_i^* in each little angle subinterval. We choose the radius of the ith pie wedge to be $r_i = f(\theta_i^*)$. The area of the ith pie wedge is given by

$$\Delta A_i = \tfrac{1}{2}(r_i)^2 \, \Delta\theta = \tfrac{1}{2}(f(\theta_i^*))^2 \, \Delta\theta$$

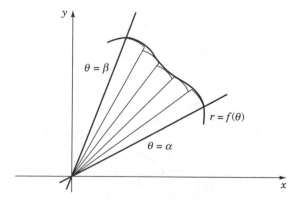

Figure 3.10 Approximating the area by pieces of pie.

So the area of the entire region is approximately equal to the sum of these, namely:

$$\sum_{i=1}^{n} \Delta A_i = \frac{1}{2} \sum_{i=1}^{n} (f(\theta_i^*))^2 \, \Delta\theta$$

If we take the limit as the wedges get thinner and thinner, then in the limit, the sum becomes an integral and we get the area:

$$A = \int_{\alpha}^{\beta} \frac{1}{2} (f(\theta))^2 \, d\theta$$

Let's try an example.

Example 1 (Seeing through the Windshield Problem)   As you are driving by the Little League fields, a large brown glob of chewing tobacco lands on your windshield. You turn on your windshield wiper and it smears the brown glob over a region on your windshield bounded by $\theta = 0$, $\theta = \pi/2$, and $r = 1 + \cos\theta$, where r is measured in feet. What is the area of that part of your windshield through which you can no longer see?

Solution   The mess appears in Figure 3.11. From there, the rest is pretty easy, huh? According to the boxed formula,

$$A = \int_{\alpha}^{\beta} \frac{1}{2} (f(\theta))^2 \, d\theta$$

$$= \int_{0}^{\pi/2} \frac{1}{2} (1 + \cos\theta)^2 \, d\theta$$

$$= \int_{0}^{\pi/2} \frac{1}{2} (1 + 2\cos\theta + (\cos\theta)^2) \, d\theta$$

$$= \int_{0}^{\pi/2} \frac{1}{2} \left(1 + 2\cos\theta + \left(\frac{1 + \cos 2\theta}{2}\right)\right) d\theta$$

$$= \frac{1}{2} \left(\theta + 2\sin\theta + \frac{\theta}{2} + \frac{\sin 2\theta}{4}\right) \Bigg|_{0}^{\pi/2}$$

$$= \frac{1}{2} \left(\frac{3\pi}{4} + 2\sin\left(\frac{\pi}{2}\right) + \frac{\sin\pi}{4}\right)$$

$$= \frac{3\pi}{8} + 1 \approx 2.178$$

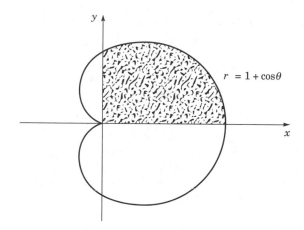

Figure 3.11 Your messy windshield.

Time to hit the car wash.

AREA BETWEEN CURVES

Many of the smears on your windshield do not reach all the way down to the origin where your wiper attaches to the car. This is lucky since the wiper always misses that part anyway. So we would like to be able to find the area between two polar curves. Suppose we want to find the area bounded by the two radial lines $\theta = \alpha$ and $\theta = \beta$ and the two curves $r = f(\theta)$ and $r = g(\theta)$, where $0 \leq f(\theta) \leq g(\theta)$ for all θ between α and β, as in Figure 3.12.

Since the area between the radial lines that is inside $r = g(\theta)$ is given by $\int_{\alpha}^{\beta} \frac{1}{2}(g(\theta))^2\, d\theta$ and the area that is inside $r = f(\theta)$ is given by $\int_{\alpha}^{\beta} \frac{1}{2}(f(\theta))^2\, d\theta$, the area that is inside $r = g(\theta)$ but outside $r = f(\theta)$ is the difference in these areas:

$$A = \int_{\alpha}^{\beta} \frac{1}{2}(g(\theta))^2\, d\theta - \int_{\alpha}^{\beta} \frac{1}{2}(f(\theta))^2\, d\theta = \frac{1}{2}\int_{\alpha}^{\beta} (g(\theta))^2 - (f(\theta))^2\, d\theta$$

Example 2   Find the area inside the circle $r = 2\cos\theta$ but outside the circle $r = 1$.

Solution

Step 1 Draw the situation. The circle $r = 1$ is easy. What about $r = 2\cos\theta$? As we did in the previous section, we can convert to rectangular coordinates:

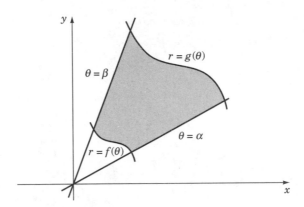

Figure 3.12 Area between $\theta = \alpha$, $\theta = \beta$, $r = g(\theta)$, and $r = f(\theta)$.

Multiply both sides of the equation by r to obtain $r^2 = 2r\cos\theta$. Then since $r^2 = x^2 + y^2$ and $x = r\cos\theta$, we can convert this equation to an equation in rectangular coordinates, $x^2 + y^2 = 2x$. Now with a little rearrangement this becomes a standard formula for a circle. We rearrange the terms to get

$$(x^2 - 2x + 1) + y^2 = 1$$
$$(x - 1)^2 + y^2 = 1$$

This is a circle of radius 1 centered at $(1, 0)$, as in Figure 3.13. We want to find the area of that crescent moon.

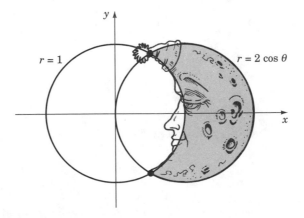

Figure 3.13 Area inside $r = 2\cos\theta$ but outside $r = 1$.

Step 2 We need to find the points of intersection of the two circles $r = 1$ and $r = 2\cos\theta$. At a point of intersection, the r-values are equal, so

$$1 = 2\cos\theta$$

$$\frac{1}{2} = \cos\theta$$

$$\theta = -\frac{\pi}{3} \text{ or } \frac{\pi}{3}$$

Step 3 Now, we can write down the integral that gives the area:

$$A = \frac{1}{2}\int_{\alpha}^{\beta}(g(\theta))^2 - (f(\theta))^2\,d\theta$$

$$= \frac{1}{2}\int_{-\pi/3}^{\pi/3}(2\cos\theta)^2 - (1)^2\,d\theta$$

$$= \frac{1}{2}\int_{-\pi/3}^{\pi/3}(4\cos^2\theta) - 1\,d\theta$$

$$= \frac{1}{2}\int_{-\pi/3}^{\pi/3}\left[4\left(\frac{1 + \cos(2\theta)}{2}\right)\right] - 1\,d\theta$$

$$= \frac{1}{2}\int_{-\pi/3}^{\pi/3}1 + 2\cos(2\theta)\,d\theta$$

$$= \frac{1}{2}\left(\theta + \sin(2\theta)\right)\Big|_{-\pi/3}^{\pi/3}$$

$$= \frac{1}{2}\left(\left[\frac{\pi}{3} + \sin\left(\frac{2\pi}{3}\right)\right] - \left[-\left(\frac{\pi}{3}\right) + \sin\left(\frac{-2\pi}{3}\right)\right]\right)$$

$$= \frac{\pi}{3} + \frac{\sqrt{3}}{2}$$

$$\approx 1.913$$

Hey, whatever rolls your socks up and down.

Infinite Series

4.1 Sequences

What's an infinite sequence?

$$1, 2, 3, 4, \ldots$$

is an infinite sequence.

So is $1, \frac{1}{2}, \frac{1}{3}, \ldots$ and $\pi, \pi^2, \pi^3, \ldots$. A *sequence* is an infinite set of real numbers that are listed in order as

$$a_1, a_2, a_3, \ldots$$

We call each of the particular numbers a *term* in the sequence. So the first term is a_1, the second term is a_2, and the n th term is a_n.

When we are in a really good mood, we might use the shorthand notation $\{a_n\}$ or $\{a_n\}_{n=1}^{\infty}$ for our sequence. So $\{1/n\}$ is a short way of writing $1, \frac{1}{2}, \frac{1}{3}, \ldots$. The sequence $\{1 + (-1)^n\}$ is just given by $0, 2, 0, 2, \ldots$.

The Sequence Hall of Fame is where the greatest sequences of all time are honored. There are the celebrities, the ones everyone knows, like 1, 2, 3, ... or 2, 4, 6, 8, ..., and there are the ones that are less well known, but revered by sequence lovers worldwide.

One of these is the *Fibonacci sequence*.

Example 1 The Fibonacci sequence, denoted F_n, is the sequence

$$1, 1, 2, 3, 5, 8, 13, 21, \ldots$$

Each term of the sequence after the first two is obtained by adding together the previous two terms. We can write a so-called recursive formula for the sequence as

$$F_1 = 1, F_2 = 1, \text{ and } F_{n+1} = F_n + F_{n-1} \qquad \text{for } n \geq 2$$

These numbers appear in a variety of places in biology, the most well-known of which is the frisky rabbit problem, but it also applies to frisky mice, frisky birds and, perish the thought, frisky porcupines.

Example 2 Rabbits are physically able to reproduce at 1 month of age, and the period for gestation is 1 month. If each time she gives birth, the female produces a pair of rabbits, who themselves will reproduce when old enough, starting with a single pair of baby male and female rabbits, how many pairs of rabbits will there be running around after n months?

Solution We begin with one pair of baby rabbits. So $F_1 = 1$. During the first month, the babies grow up and fall in love. They tie the knot at the end of the first month, when $F_2 = 1$, and shortly thereafter the female begins to crave strange flavors of ice cream. At the beginning of the third month, she gives birth to a new pair of rabbits.

So

$$F_1 = 1, F_2 = 1, F_3 = 2$$

At the beginning of the fourth month, the female gives birth to a second pair of rabbits. No slowing down for her. But her first offspring are just now old enough to consider a family. (No incest taboos in this family.) So $F_4 = 3$. At the beginning of the fifth month, both the first mother and her first daughter give birth to a pair of rabbits each, and there is quite a celebration, with carrot juice, and lots of hopping around. So $F_4 = 5$, and it's time to expand the rabbit hutch. As you can see, at the beginning of month $n + 1$, the total number of pairs of rabbits will be the total number from the previous month, which was F_n, plus the number of new pairs born to rabbit pairs old enough to breed, which is F_{n-1}.

Hey, whatever makes your nose twitch.

4.2 Limits of sequences

What happens when you follow a sequence really far out, right to the limit? Let's take a look at limits of sequences, like $\lim\limits_{n \to \infty} \dfrac{1}{n} = 0$ or $\lim\limits_{n \to \infty} n = \infty$. The idea is to take the limit of the function $f(n)$ just like we took limits of functions $f(x)$ as x went to ∞. All of the standard rules apply.

Example   Determine the limit of the sequence $\left\{ \dfrac{\ln n}{n} \right\}$.

Solution   We treat this sequence just the way we would treat the function $\dfrac{\ln x}{x}$ if we wanted to take its limit as x went to ∞. Letting n go to ∞, we see that this expression is of the form ∞/∞. So it's an indeterminate form, which means we get to apply L'Hôpital's rule:

$$\lim_{n \to \infty} \frac{\ln n}{n} = \lim_{n \to \infty} \frac{1/n}{1} = 0$$

Hey, whatever pops your bubble wrap.

4.3 Series: the basic idea

What's the difference between a good series and a bad series? In a good series, the plot lines all converge at the end of each episode, and you have some sense of completion. Examples include *Seinfeld, MASH, Gilligan's Island*, and *Murphy Brown*. In a bad series, the various plotlines meander this way and that, never converging at the end of the show, and leaving the viewers confused and depleted. Examples include *Melrose Place, The Young and the Restless*, and *Buffy the Vampire Slayer*.

So it's clear that in order to have a successful series, you need convergence. Well, that is the essense of this chapter. But the series here don't involve Elaine's troubles with the Soup Nazi. Instead, they involve summing up a whole lot of numbers, and we mean a WHOLE LOT. We will take all of the numbers in an infinite sequence and add them together.

An **infinite series** is a sum of the form:

$$a_1 + a_2 + a_3 + \cdots + a_n + \cdots$$

We will also express this in the summation notation as $\sum\limits_{n=1}^{\infty} a_n$. Sometimes we start with $n = 0$, so

$$b_0 + b_1 + b_2 + \cdots + b_n + \cdots = \sum_{n=0}^{\infty} b_n$$

But we are going to have to be careful about what an infinite sum like this means. How do we make sense of adding up infinitely many numbers? As we usually do when confronted with infinity in math, we define it in terms of a limit. But first, let's look at the corresponding finite sums.

Definition The *nth partial sum* of a series is given by

$$S_n = a_1 + a_2 + \cdots + a_n$$

In other words, it is the sum of the first n terms of the series.

$$S_1 = a_1$$
$$S_2 = a_1 + a_2$$
$$S_3 = a_1 + a_2 + a_3$$

As we continue, S_n is getting to look more and more like the infinite series. So it makes sense to say:

Sum of a Series If $S = \lim\limits_{n \to \infty} S_n$, then

$$\sum_{n=1}^{\infty} a_n = S$$

We say the series *converges* to S. If $\lim\limits_{n \to \infty} S_n$ does not exist, we say that the series *diverges*.

Example 1 $1 + 1 + 1 + \cdots$. Notice that $S_1 = 1, S_2, = 2$, and $S_n = n$. So

$$\lim_{n \to \infty} S_n = \lim_{n \to \infty} n = \infty$$

The limit does not exist and the series diverges.

Example 2 $\frac{1}{2} + \frac{1}{4} + \frac{1}{8} + \cdots$

Then $S_1 = \frac{1}{2}, S_2 = \frac{3}{4}, S_3 = \frac{7}{8}$, and $S_n = \dfrac{2^n - 1}{2^n}$. Since

$$S = \lim_{n \to \infty} S_n = \lim_{n \to \infty} \frac{2^n - 1}{2^n} = \lim_{n \to \infty} \frac{1 - 1/2^n}{1} = 1$$

we know that the series converges to 1.

These two examples demonstrate the two basic behaviors of series. Some blow up in your face and some converge nicely to a finite number. One way to think about this basic difference is in terms of bathtubs.

THE BATHTUB ANALOGY

Think of an infinite series in terms of putting water in a bathtub. We begin by plugging up the empty tub. Then we use a measuring cup to dump in an amount of water corresponding to the first term in the series. We then dump in an amount of water corresponding to the second term. We repeat this process with each subsequent term in the series, as in Figure 4.1. So in Example 1 above, we put in first 1 cup of water, and then another cup and another cup. Eventually the bathtub is going to overflow. It doesn't matter how big the bathtub is, eventually it will overflow. It could be as big as the Pacific Ocean and eventually it will overflow. It could be as big as Donald Trump's ego and eventually it will overflow. That's what it means to diverge.

On the other hand, if we took the second series, the story is a bit different. First we put a half cup of water in the tub, then a quarter cup, and then an eighth of a cup. We continue this process forever. Does the tub overflow? Hardly. After repeating this process an infinite number of times, we have a grand total of 1 cup of water in the tub. Not enough to wash a big hamster.

Notice that the bathtub analogy tells us something else. The first few terms are really irrelevant when deciding whether a series converges or diverges. If the series does diverge, it will overflow any size bathtub, and it won't matter if the first term was a cup or half a cup. In fact, now that you

Figure 4.1 Summing a series is like filling a bathtub.

mention it, it doesn't matter what the first million terms look like. Ultimately, what determines convergence or divergence is the rest of the terms, not the first million or billion. So the series

$$1 + 1 + 1 + 1 + 1 + 1 + 1 + 1 + 1 + 1 + \tfrac{1}{2} + \tfrac{1}{4} + \tfrac{1}{8} + \tfrac{1}{16} + \cdots$$

converges since the series

$$\tfrac{1}{2} + \tfrac{1}{4} + \tfrac{1}{8} + \tfrac{1}{16} + \cdots$$

converges. The same holds even if there were a trillion 1's at the beginning of the first series.

Hey, whatever squares your root.

4.4 Geometric series: the extroverts

When confronted with a series, the goal is to determine whether or not it converges, and if it does, to try and find out to what. But series do not readily disclose this information. Many of them prefer not to share.

Luckily, many series do not fall in this category. They are more than happy to tell you whether or not they converge, and if they converge, to what. Given the chance, they will talk your ear off. They are the friendliest series you are likely to meet, true extroverts.

Speaking of which, how do you tell an introverted mathematician from an extroverted mathematician?

The extrovert looks at *your* shoes.

Back to the extroverted series. Here are some examples:

$$1 + \tfrac{1}{2} + \tfrac{1}{4} + \tfrac{1}{8} + \cdots$$
$$1 + \tfrac{1}{3} + \tfrac{1}{9} + \tfrac{1}{27} + \cdots$$

These are *geometric series*, as the terms form what is called a *geometric progression*. This means that each term is obtained from the previous term by multiplying by a fixed number. For the first series, the fixed number is $\tfrac{1}{2}$, and for the second series, the fixed number is $\tfrac{1}{3}$. The general form of a series like this is

$$a + ar + ar^2 + ar^3 + \cdots$$

Every term is obtained from the previous term by multiplying by the same number r. When you come across a geometric series, it's a cause for celebration, because geometric series are completely understood. So pull out the party hats!

Geometric Series Test The geometric series $a + ar + ar^2 + ar^3 + \cdots$ converges to $\dfrac{a}{1-r}$ if $|r| < 1$ and diverges if $|r| \geq 1$.

We won't go through the argument for this, but it's not hard, so check to see if your professor expects you to know it.

Let's apply it.

Example 1 $\tfrac{1}{3} + \tfrac{1}{9} + \tfrac{1}{27} + \cdots$

Solution The series is geometric with $a = \tfrac{1}{3}$ and $r = \tfrac{1}{3} \leq 1$, so it converges to $\dfrac{a}{1-r} = \dfrac{1/3}{1 - 1/3} = \tfrac{1}{2}$.

Example 2   A ball of flubber is dropped from 6 feet, and bounces to a height of two-thirds its previous height each time it bounces. How much vertical distance up and down does it travel? (See Figure 4.2.)

Solution We see that the flubber travels 6 feet in its first drop and then both up and down a distance $(2/3)6 = 4$ feet, and then up and down a distance $(2/3)^2 6 = 4(2/3)$ feet, etc. So the flubber travels a total distance

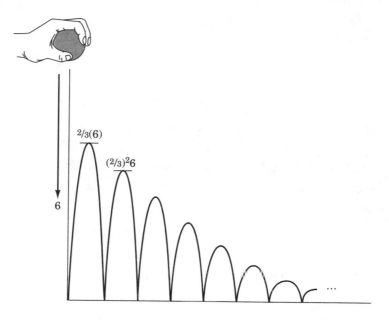

Figure 4.2 Finding the total vertical distance traveled by bouncing flubber.

$$S = 6 + 2(4) + 2(4)(2/3) + 2(4)(2/3)^2 + \cdots = 6 + 8 + 8(2/3) + 8(2/3)^2 + \cdots$$

Temporarily dropping the first term, this is a geometric series with $a = 8$ and $r = 2/3$. So

$$S = 6 + \frac{8}{1 - 2/3} = 6 + 24 = 30 \text{ feet}$$

Hey, whatever bounces your superball.

4.5 *n*th-term test

This is the most basic test to apply to a series if you are trying to decide whether or not it converges. If the series doesn't pass this test, it's over. The series diverges. This test says that if the individual terms are not getting really small, then the bathtub will overflow. In mathematical language:

***n*th-Term Test**   If $\lim_{n \to \infty} a_n \neq 0$, the infinite series $\sum_{n=1}^{\infty} a_n$ diverges.

Example 1 $\frac{1}{2} + \frac{2}{3} + \frac{3}{4} + \cdots + \dfrac{n}{n+1} + \cdots.$

Solution   This series diverges since $\lim\limits_{n \to \infty} \dfrac{n}{n+1} = 1 \neq 0.$

Talk about an easy test to perform. BUT THIS TEST DOES NOT WORK IN REVERSE. There are series where $\lim\limits_{n \to \infty} a_n = 0$, but the series still does not converge. The classic example is the so-called harmonic series given by $1 + \frac{1}{2} + \frac{1}{3} + \cdots$. We'll shortly see that it diverges.

Common Mistake   Just because the a_n's shrink to 0 does not mean that the series $\sum a_n$ converges.

4.6 Integral test and p-series: more friends

We would like to know whether or not a series converges. Does it add up to some finite number or does it blow up in your face? We will develop a collection of tests to see what it does. Some of the tests will show that the particular series we're looking at and another series are partners in crime, meaning the one converges if and only if the other one does. This next test shows that under certain circumstances, the series and an improper integral are partners in crime.

This test and the next few will apply to *positive term series,* that is, series in which all the terms are positive numbers.

Integral Test   If $a_n = f(n)$, where f is positive valued, continuous, and decreasing for $x \geq 1$, then $\displaystyle\int_1^\infty f(x)\,dx$ and $\displaystyle\sum_{n=1}^\infty a_n$ are partners in crime; either both converge or both diverge.

Why It Works   We will first show that if $\displaystyle\int_1^\infty f(x)\,dx$ converges, then $\displaystyle\sum_{n=1}^\infty a_n$ should converge as well. From Figure 4.3, we see that the area under the curve $y = f(x)$, as x goes from 1 to ∞, is greater than the sum of the areas under the rectangles. But notice that the first rectangle has height $f(2)$, which is a_2, and width 1. So the first rectangle has area a_2. The second rectangle has height $f(3)$, which is a_3, and width 1. So it has area a_3. As we continue with the rectangles, we see that the nth rectangle has area a_{n+1}.

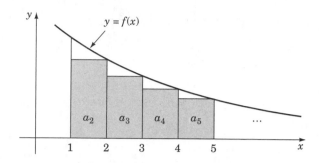

Figure 4.3 $\int_1^\infty f(x)\,dx > a_2 + a_3 + a_4 + \cdots.$

So the sum of the areas of all the rectangles is $a_2 + a_3 + a_4 + \cdots$. Since this area is less than the area given by $\int_1^\infty f(x)\,dx$, we have

$$\int_1^\infty f(x)\,dx > a_2 + a_3 + a_4 + \cdots.$$

But we're assuming the larger area $\int_1^\infty f(x)\,dx$ is finite. So the smaller area must also be finite. (There is actually more work to do here, since just having an upper bound does not ensure convergence, but we won't go into that.)

So, $a_2 + a_3 + a_4 + \cdots$ converges. But if it converges, adding one term a_1 won't change that, so $a_1 + a_2 + a_3 + a_4 + \cdots$ converges as well.

Now we'll show that if $\int_1^\infty f(x)\,dx$ diverges, then so does the series. We'll work from Figure 4.4. Here, the sum of the areas inside the rectangles will be greater than the area under the curve. The area in the first rectangle is a_1, in the second rectangle a_2, etc. So we see that

$$a_1 + a_2 + a_3 + \cdots > \int_1^\infty f(x)\,dx$$

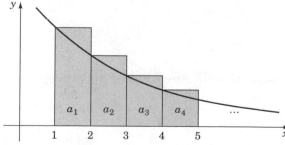

Figure 4.4 $\int_1^\infty f(x)\,dx < a_1 + a_2 + a_3 + \cdots.$

But $\displaystyle\int_1^\infty f(x)\,dx = \infty$, so the larger area must also be ∞. Therefore the series diverges.

HARMONIC SERIES

Let's apply the integral test to the harmonic series given by

$$1 + \tfrac{1}{2} + \tfrac{1}{3} + \tfrac{1}{4} + \cdots = \sum_{n=1}^{\infty} \frac{1}{n}$$

This is possibly the most famous series alive. Why, you ask? What makes a series famous? What makes anyone famous? First you have to get a little famous, and then you have to be involved in a scandal. Well, $1 + \tfrac{1}{2} + \tfrac{1}{3} + \cdots$ is obviously a little famous. After all, we just take the reciprocals of 1, 2, 3, 4, ... and add them together. It's hard to imagine a simpler series that satisfies the nth term test and that therefore has a chance to converge.

And what's the scandal? Let's see. Since $a_n = 1/n = f(n)$ where $f(x) = 1/x$ and since $1/x$ is positive, continuous and decreasing for all $x \geq 1$, the integral test applies.

So $\displaystyle\sum_{n=1}^{\infty} \frac{1}{n}$ and $\displaystyle\int_1^\infty \frac{1}{x}\,dx$ are partners in crime. They either both converge or both diverge. But

$$\int_1^\infty \frac{1}{x}\,dx = \lim_{b\to\infty} \ln x \,\Big|_1^b = \infty$$

So the integral and the series both diverge! The tiny terms in this series, $1, \tfrac{1}{2}, \tfrac{1}{3}, \ldots$, add up to infinity!

In other words, if we start with a bathtub, and we put in 1 cup of water, and then half a cup and then a third, eventually it will overflow. If we start with an empty Pacific Ocean, and put in 1 cup of water, then a half and then a third, eventually it will overflow.

But it will take a long, long time. Just to have enough terms to add up to 5, we must use 83 terms in the series. To add to 10, we need to use 12,667 terms in the series. To add to 20, we need about 272,400,000 terms in the series. And to add to 1000, we must use 1.1×10^{434} terms in the series. To put this in perspective, it is estimated that there are 10^{80} atoms in the entire observable universe. This is a series that diverges, but oh so very slowly. Don't use this to fill your bathtub, unless you have a few quadrillion years on your hands and nothing better to do.

p-SERIES

The harmonic series is part of a family of series well known to people who hang out with series. We could only be speaking of the famous p-series.

Definition   A *p-series* is a series of the form $\sum\limits_{n=1}^{\infty} \dfrac{1}{n^p}$, with $p > 0$.

Notice that if $p \leq 0$, the series would diverge by the nth term test.

We really like p-series because they are not shy. They are happy to tell you if they converge or not, and that's on the first date.

---

p-Series Test   A p-series $\sum\limits_{n=1}^{\infty} \dfrac{1}{n^p}$ converges for $p > 1$ and diverges for $p \leq 1$.

Why It Works   The test works for $p = 1$, since this is the harmonic series and we have already checked that case. So we can assume that $p \neq 1$. We can also assume $p > 0$ since otherwise the terms do not get small. We will apply the integral test with $a_n = 1/n^p = f(n)$ and $f(x) = 1/x^p$.

Since $1/x^p$ is a positive, continuous, and decreasing function for $x \geq 1$, we can use the integral test:

$$\int_1^{\infty} \frac{1}{x^p}\, dx = \frac{x^{-p+1}}{-p+1}\bigg|_1^{\infty}$$

When $p > 1$, the exponent $(-p + 1)$ is negative. So it's as if the x is in the denominator with positive exponent $-(-p + 1) = p - 1$, and when we evaluate $\dfrac{1}{x^{p-1}(1 - p)}$ as $x \to \infty$, we get 0. When $p < 1$, the exponent $(-p + 1)$ is positive, so when we evaluate the limit as $x \to \infty$, we get ∞.

So the integral and the series converge for $p > 1$ and diverge for $p \leq 1$.

Examples

1. $\sum\limits_{n=1}^{\infty} \dfrac{1}{n^3}$ converges since $p = 3 > 1$.

2. $\sum\limits_{n=1}^{\infty} \dfrac{1}{\sqrt{n}}$ diverges since $p = \frac{1}{2} \leq 1$.

Hey, whatever climbs your beanstalk.

4.7 Comparison tests

The name's Marley. Philip Marley. And this is my associate and cousin, Bob. That's right, Bob Marley. We're investigators. Yup, you heard right. Investigators for the Invigor Health Maintenance Organization. Perhaps you've heard of us. We've been in the press a lot lately for giving free trips to Aruba to doctors who minimize services. "That leg's not broken. It's always looked like that." That kind of thing.

And I know what you are thinking. What moral individual would work for a company like this? The answer is, "No moral individual." I guess that tells you something about me. And in a parasitic organization like this one, who are the lowest of the low, the bottom-feeders, the ones who ultimately determine whether you get the Viagra or not? That's right, it's me and Bob. The Marleys. Our job is to investigate claims for payment. This particular case involved an especially insidious situation. You've heard about it, but to most people, that's as close as they ever come. That's right, I'm talking about domination. Domination of one series by another. Seems that when each term in a series is at least as large as the corresponding term in a second series, we say the first series dominates the second. That is to say, if $\sum a_i$ and $\sum b_i$ are the two series and for every single possible i, $a_i \geq b_i$, then we say $\sum a_i$ dominates $\sum b_i$.

In case you're not keeping up, take $\sum \dfrac{1}{n}$ and $\sum \dfrac{1}{n^2}$. Then

$$\frac{1}{n} \geq \frac{1}{n^2} \quad \text{for } n = 1, 2, 3, \ldots$$

so $\sum \dfrac{1}{n}$ dominates $\sum \dfrac{1}{n^2}$. Get the idea? Not pretty, is it? But some series get off on humiliation. What can you do?

Now why should anybody care about such things? If two series want to play a little domination in private, should it be any skin off my nose, or Bob's? No, of course not. Live and let live. Whatever trims your shrubbery. But this particular pair of series likes it rough. Real rough. We're talking about a stilletto heel firmly planted in the center of the back, and it's all the better if you have a slipped disk.

And of course, that's where we come in. The company needs to know the medical costs incurred by domination. What are the risks? What are the dangers? Are we going to be picking up the tab for 6 years of recuperative physical therapy? That's where the basic comparison tests come in. They show the implications.

> **Basic Comparison Test**   If $\sum_{n=1}^{\infty} a_n$ and $\sum_{n=1}^{\infty} b_n$ are positive term series, and
> if $b_n \geq a_n$ for all n, then:
>
> **1.** If $\sum_{n=1}^{\infty} b_n$ converges, then so does $\sum_{n=1}^{\infty} a_n$.
>
> **2.** If $\sum_{n=1}^{\infty} a_n$ diverges, then so does $\sum_{n=1}^{\infty} b_n$.

This only holds if all the terms of the series are positive numbers. To paraphrase the result,

"If the bigger series converges, so does the smaller series."

and

"If the smaller series diverges, so does the bigger series."

This can get confusing, so Bob and I have found that the easiest way to keep it straight is in terms of balloons.

THE BALLOON ANALOGY

Think of adding up the terms in an infinite series as blowing air into a balloon, as in Figure 4.5. Each term adds some additional air to the balloon. If you add too much air, then the balloon pops, and the series diverges. If the balloon doesn't pop, then the series converges to a finite number which is the volume of air in the balloon.

We can picture the basic comparison test in terms of two balloons, one inside the other, as in Figure 4.6. Let's say the one on the inside represents $\sum a_n$ and the one on the outside represents $\sum b_n$. So we are assuming throughout that $\sum b_n$ dominates $\sum a_n$, meaning each term $b_n \geq a_n$.

But think about balloons for a second. If the outside balloon doesn't pop, meaning that the series $\sum b_n$ converges, and the volume of the balloon is finite, the inside balloon can't pop either, since it has a smaller volume. So the smaller series, $\sum a_n$, also converges.

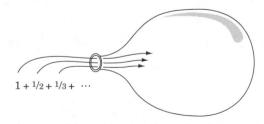

$1 + \frac{1}{2} + \frac{1}{3} + \cdots$

Figure 4.5 Each term in the series adds air to the balloon.

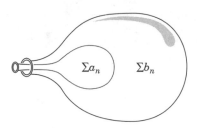

Figure 4.6 $\sum_{n=1}^{\infty} b_n$ dominates $\sum_{n=1}^{\infty} a_n$.

On the other hand, if the smaller balloon does pop, meaning the smaller series $\sum a_n$ diverged, then the bigger balloon must also have popped, since it must have a larger volume, meaning the bigger series $\sum b_n$ must also have diverged. Simple enough.

But there are two other possibilities as well. What do they say? If the smaller balloon doesn't pop, the bigger balloon may not pop or it may pop. We get absolutely no info. No help here. Nada informata. And if the bigger balloon pops, so what? The smaller balloon is completely unaffected by that. It might pop, it might not. The balloon analogy allows us to keep track of the four possibilities, two of which tell us something and two of which do not.

Example 1 Determine which of the following series converge or diverge.

a. $\sum_{n=1}^{\infty} \dfrac{1}{2 + 3^n}$

$\dfrac{1}{2 + 3^n} \le \dfrac{1}{3^n}$ since $\dfrac{1}{2 + 3^n}$ has the bigger denominator. So

$$\sum_{n=1}^{\infty} \dfrac{1}{3^n} \text{ dominates } \sum_{n=1}^{\infty} \dfrac{1}{2 + 3^n}$$

The bigger series $\sum_{n=1}^{\infty} \dfrac{1}{3^n}$ is a geometric series with $r = \frac{1}{3} < 1$, so it converges. The smaller series $\sum_{n=1}^{\infty} \dfrac{1}{2 + 3^n}$ converges as well, by the basic comparison test. Big balloon doesn't pop implies small balloon doesn't pop.

b. $\sum_{n=1}^{\infty} \dfrac{1}{\sqrt{n} - 1}$

$\dfrac{1}{\sqrt{n} - 1} > \dfrac{1}{\sqrt{n}}$ because $\dfrac{1}{\sqrt{n} - 1}$ has the smaller denominator.

Since the smaller series is $\sum_{n=1}^{\infty} \dfrac{1}{\sqrt{n}}$, a p-series with $p = \frac{1}{2} < 1$, it diverges. By the basic comparison test, if the smaller series diverges, so must the larger. Smaller balloon pops means bigger balloon pops.

c. $\displaystyle\sum_{n=1}^{\infty} \dfrac{1}{n+1}$

Our first inclination is to compare this to $\dfrac{1}{n}$. But $\dfrac{1}{n+1} < \dfrac{1}{n}$, and $\displaystyle\sum_{n=1}^{\infty} \dfrac{1}{n}$ diverges since it is the harmonic series. If the larger series diverges (big balloon pops), then we know absolutely nothing about the smaller series. (Small balloon could pop, or not.)

So this comparison was a waste of time. Too bad. What else could we do? How about this?

$$n + 1 \leq n + n = 2n, \text{ so } \dfrac{1}{n+1} \geq \dfrac{1}{2n}. \text{ But the smaller series is}$$

$$\sum_{n=1}^{\infty} \dfrac{1}{2n} = \dfrac{1}{2} \sum_{n=1}^{\infty} \dfrac{1}{n}$$

This is just one-half times the harmonic series, and so it diverges. Half of infinity is still infinity! And if the smaller series diverges, so does the larger. Done!

Finally, we will just note that in fact there is an easier way to do this last one, which is just to write out the series.

$$\sum_{n=1}^{\infty} \dfrac{1}{n+1} = \frac{1}{2} + \frac{1}{3} + \frac{1}{4} + \cdots$$

We see that this series is really just the harmonic series missing its first term, so it must diverge.

d. $\displaystyle\sum_{n=1}^{\infty} \dfrac{1}{\sqrt{n}+1}$

This one is a problem. Notice that if we try to apply the basic comparison test in the obvious way, we have $\dfrac{1}{\sqrt{n}+1} < \dfrac{1}{\sqrt{n}}$.

Unfortunately, the bigger series is a p-series with $p = \frac{1}{2} < 1$, so it diverges, telling us nothing about the smaller series. We need a new test! What luck! Here comes one now.

Limit Comparison Test If $\displaystyle\sum_{n=1}^{\infty} a_n$ and $\displaystyle\sum_{n=1}^{\infty} b_n$ are positive term series, and for some positive number k,

$$\lim_{n \to \infty} \frac{a_n}{b_n} = k$$

then either both series converge or both series diverge.

They are partners in crime, the Bonnie and Clyde of the series set. The point is that the two series are behaving very similarly when n gets large. If $k = 1$, the terms in the two series are getting closer and closer to being the same. If k equals any other number, the terms are getting closer and closer to being just a multiple by k of each other. So we expect the series to behave the same. It is only when the limit is 0 or ∞ that the series might not behave the same.

Example 2. Does $\displaystyle\sum_{n=1}^{\infty} \frac{1}{\sqrt{n} + 1}$ converge?

Solution We'll let $a_n = \dfrac{1}{\sqrt{n} + 1}$. We tried the basic comparison test on this already but it didn't work.

So we'll try the limit comparison test instead. Let's compare with the series $b_n = \dfrac{1}{\sqrt{n}}$. Then

$$\lim_{n \to \infty} \frac{a_n}{b_n} = \lim_{n \to \infty} \frac{\dfrac{1}{\sqrt{n}+1}}{\dfrac{1}{\sqrt{n}}} = \lim_{n \to \infty} \frac{\sqrt{n}}{\sqrt{n}+1}$$

$$= \lim_{n \to \infty} \frac{1}{1 + \dfrac{1}{\sqrt{n}}} = 1$$

So these two series are partners in crime. What one does the other does. One for all and all for one. Since $\displaystyle\sum_{n=1}^{\infty} \frac{1}{\sqrt{n}}$ is a p-series with $p = \frac{1}{2}$, it diverges. So our original series diverges as well.

Hey, whatever dunks your doughnuts.

4.8 Alternating series and absolute convergence

A series of the form

$$a_1 - a_2 + a_3 - a_4 + \cdots + (-1)^{n+1}a_n + \cdots$$

or of the form

$$-a_1 + a_2 - a_3 + a_4 + \cdots + (-1)^n a_n + \cdots$$

is called an *alternating series,* where we're assuming that each a_n is positive. When we start with a series of all positive terms, and change the sign of every other term, we get an alternating series. For example, we might have

$$1 - \tfrac{1}{2} + \tfrac{1}{3} - \tfrac{1}{4} + \tfrac{1}{5} - \cdots$$

Often, it is relatively easy to tell that such a series converges.

Alternating Series Test   If each a_n is positive, $a_{n+1} \leq a_n$ for all n, and $\lim\limits_{n \to \infty} a_n = 0$, then the corresponding alternating series converges.

Let's see how we apply this.

Example 1   Does $\displaystyle\sum_{n=1}^{\infty}(-1)^{n+1}\frac{1}{n}$ converge?

Solution   First we check if the first condition is satisfied. We look at the sequence of positive terms $a_1 = 1, a_2 = \tfrac{1}{2}, \ldots, a_n = 1/n$. Is it true that

$$a_{n+1} \leq a_n?$$

$$\frac{1}{n+1} \leq \frac{1}{n}?$$

Yup; $\dfrac{1}{n+1}$ has the bigger denominator.

And we check the second condition:

$$\lim_{n \to \infty} \frac{1}{n} = 0$$

So the alternating series test is satisfied, and the series converges. Note that this series is the alternating version of the harmonic series, which diverged. Just like a pair of siblings, one of whom turns out kind and sweet while the other is evil incarnate, the alternating harmonic series converges while the harmonic series blows up in your face.

Example 2   Does $\sum_{n=1}^{\infty} (-1)^{n+1} \dfrac{n}{n^2 + 1}$ converge?

Solution   Now we take $a_n = \dfrac{n}{n^2 + 1}$. We check the first condition of the alternating series test. Is it true that

$$a_{n+1} \le a_n?$$

$$\frac{n + 1}{(n + 1)^2 + 1} \le \frac{n}{n^2 + 1}?$$

$$(n^2 + 1)(n + 1) \le ((n + 1)^2 + 1)n?$$

$$n^3 + n^2 + n + 1 \le n^3 + 2n^2 + 2n?$$

$$0 \le n^2 + n - 1?$$

This is certainly true for all $n \ge 1$.

Before we go on to check the second condition, let's take a look at a second method for showing that $a_{n+1} \le a_n$ in this particular example. We can demonstrate that the function $f(x) = \dfrac{x}{x^2 + 1}$ is decreasing. Then it will be the case that $f(n + 1) \le f(n)$, which is the same as $a_{n+1} \le a_n$. We can show that $f(x) = \dfrac{x}{x^2 + 1}$ is decreasing by showing that its derivative is negative. But

$$f'(x) = \frac{1(x^2 + 1) - 2x(x)}{(x^2 + 1)^2} = \frac{1 - x^2}{(x^2 + 1)^2} < 0$$

for all $x > 1$.

And of course, we can't forget to check the second condition:

$$\lim_{n \to \infty} \frac{n}{n^2 + 1} = \lim_{n \to \infty} \frac{1}{2n} = 0$$

by L'Hôpital's rule.

So this series converges, too. Hey, whatever skims your milk.

ABSOLUTE CONVERGENCE

When someone is crazy, we call him *nuts*. When someone is loonier than a dentist on nitrous oxide, we call him *absolutely nuts*.

When the terms of a series add up to a finite quantity, we call it *convergent*. But when the absolute values of its terms adds up to a finite quantity, we call it *absolutely convergent*. Absolute convergence is a stronger form of convergence.

> **Definition** A series $\sum_{n=1}^{\infty} a_n$ is *absolutely convergent* if the series $\sum_{n=1}^{\infty} |a_n|$ converges.

For instance, $\sum_{n=1}^{\infty} \frac{1}{n^2}$ is convergent and absolutely convergent, since $\sum_{n=1}^{\infty} \frac{1}{n^2}$ is a p-series with $p > 1$ and $1/n^2$ is its own absolute value.

On the other hand, $\sum_{n=1}^{\infty} \frac{(-1)^n}{n}$ is convergent but not absolutely convergent, as its absolute value gives the harmonic series. We call such a series *conditionally convergent*.

Notice that if someone is absolutely nuts, then they are certainly nuts. The same holds true for convergence.

> **Absolute Convergence Theorem** If a series is absolutely convergent, it is convergent.

In the case of a positive term series, this theorem says absolutely nothing, since the terms of the series don't change when you take absolute values. But sometimes it's quite useful.

Example 3 The series $\sum_{n=1}^{\infty} \frac{(-1)^n}{n^2}$ is absolutely convergent, because taking the absolute value of its terms gives the series $\sum_{n=1}^{\infty} \frac{1}{n^2}$, which is a p-series with $p = 2 > 1$, and this series converges by the p-series test. So the original series is absolutely convergent and therefore convergent. Notice that we could have used the alternating series test to show it converged, but hey, why bother? We have better things to do.

4.9 More tests for convergence

A man carries his dog into the vet's office and says, "Doctor, you have to help me. My dog's not well." The vet lays the dog down on an examination table and puts a stethoscope to the dog's chest. He sighs as he straightens up and says, "I'm sorry, sir, but your dog is dead."

"That can't be," says the man, completely distraught. "I want a second opinion."

"Okay," shrugs the vet. He leaves the room only to return with a large tabby cat, which he sets down next to the dog on the examining table. The cat carefully sniffs the dog from head to tail and then meows at the vet. The vet says, "The cat has confirmed it. The dog is dead."

"No," says the man, desperate in his grief. "You are both wrong."

"Okay," says the vet, and he goes out again, returning with a black labrador retriever. The lab puts his forepaws up on the examining table and nudges the dog with his nose several times, but the dog doesn't budge. Then the lab barks at the vet.

"Same diagnosis, I am afraid," says the vet. Sobbing, the man says, "Okay, okay, I believe you. What do I owe you?"

"That'll be $550," says the vet.

"What?" says the man. "That's outrageous. Why so much?"

"Well, look," says the vet, "it's $50 for my diagnosis. Then it's $250 for the cat scan and $250 for the lab test."

Speaking of tests, the next one is perhaps the most commonly applied of them all.

Ratio Test If $\displaystyle\sum_{n=1}^{\infty} u_n$ is a series, then

1. If $\displaystyle\lim_{n \to \infty} \left| \frac{u_{n+1}}{u_n} \right| < 1$, the series converges absolutely.

2. If $\displaystyle\lim_{n \to \infty} \left| \frac{u_{n+1}}{u_n} \right| > 1$, the series diverges.

3. If $\displaystyle\lim_{n \to \infty} \left| \frac{u_{n+1}}{u_n} \right| = 1$, then we know squat. It could converge or diverge, but this test sure isn't going to tell us.

Let's do an example to see how this works.

Example 1   Test $\displaystyle\sum_{n=1}^{\infty} \frac{(-1)^n n}{3^n}$ for convergence.

Solution   In this case, $u_n = \dfrac{(-1)^n n}{3^n}$. We get u_{n+1} from u_n by replacing each n with an $n + 1$. So

$$u_{n+1} = \frac{(-1)^{n+1}(n + 1)}{3^{n+1}}$$

Then

$$\lim_{n \to \infty}\left|\frac{u_{n+1}}{u_n}\right| = \lim_{n \to \infty}\left|\frac{\dfrac{(-1)^{n+1}(n + 1)}{3^{n+1}}}{\dfrac{(-1)^n n}{3^n}}\right|$$

$$= \lim_{n \to \infty}\left|\frac{n + 1}{3n}\right|$$

$$= \lim_{n \to \infty}\frac{1 + 1/n}{3}$$

$$= \frac{1}{3} < 1$$

By the ratio test, this is an absolutely convergent series, and therefore convergent.

Here's another test that comes in handy once in a while.

Root Test   If $\displaystyle\sum_{n=1}^{\infty} u_n$ is a series, then

1. If $\displaystyle\lim_{n \to \infty}\left(|u_n|\right)^{1/n} < 1$, the series converges absolutely.

2. If $\displaystyle\lim_{n \to \infty}\left(|u_n|\right)^{1/n} > 1$, the series diverges.

3. If $\displaystyle\lim_{n \to \infty}\left(|u_n|\right)^{1/n} = 1$, then we know squat. It could converge or diverge, but this test sure isn't going to tell us.

Example 2 Test $\sum_{n=1}^{\infty} \dfrac{1}{e^n}$ for convergence.

Solution

$$\lim_{n \to \infty} \left| \frac{1}{e^n} \right|^{1/n} = \lim_{n \to \infty} \left| \frac{1}{e} \right|$$

$$= \frac{1}{e}$$

$$= \frac{1}{2.718\ldots} < 1$$

By the root test, this is an absolutely convergent series, and therefore convergent. If you really want to pound this one into the ground, you can also do it by the ratio test. And yes, we know, this is also a geometric series with $r = 1/e$, so it converges for that reason, too. But don't rub it in.

Whatever braises your chicken.

  ## Power series

A *power series* is a series of the form

$$\sum_{n=0}^{\infty} a_n x^n = a_0 + a_1 x + a_2 x^2 + \cdots + a_n x^n + \cdots$$

These series behave like firehoses. If kept under control, they can put out fires and disperse students rioting over low grades. But you don't want to lose your grip on a firehose. If you do, run for cover, because they start flipping around like an eel plugged into an electrical outlet.

In the case of power series, it is the values that we put in for x that determine whether or not the series is under control. Some values of x will cause the series to converge. But put in a wrong value and the series diverges, meaning it's a firehose out of control. Run for cover. We call the values of x for which a power series converges its *interval of convergence*.

Let's look at some examples.

Example 1 Determine the values of x for which the following series converges.

$$\sum_{n=0}^{\infty} \frac{x^n}{n!} = 1 + x + \frac{x^2}{2!} + \frac{x^3}{3!} + \cdots$$

Solution We apply the ratio test:

$$\lim_{n \to \infty} \left| \frac{u_{n+1}}{u_n} \right| = \lim_{n \to \infty} \left| \frac{\dfrac{x^{n+1}}{(n+1)!}}{\dfrac{x^n}{n!}} \right|$$

$$= \lim_{n \to \infty} \frac{|x|}{n+1} \qquad \text{(For any fixed } x \text{, this goes to 0.)}$$

$$= 0$$

It doesn't matter what x is. The series converges for all values of x.

Example 2 Determine the values of x for which the following series converges.

$$\sum_{n=1}^{\infty} \frac{nx^n}{3^n} = \frac{x}{3} + \frac{2x^2}{9} + \frac{3x^3}{27} + \cdots$$

Solution We apply the ratio test again:

$$\lim_{n \to \infty} \left| \frac{u_{n+1}}{u_n} \right| = \lim_{n \to \infty} \left| \frac{\dfrac{(n+1)x^{n+1}}{3^{n+1}}}{\dfrac{nx^n}{3^n}} \right| = \lim_{n \to \infty} \frac{|x|(n+1)}{3n}$$

$$= \frac{|x|}{3}$$

By the ratio test, if $\dfrac{|x|}{3} < 1$, the series will converge. So $|x| < 3$ gives convergence.

If $-3 < x < 3$, the series definitely converges. For values of $x < -3$ and $x > 3$, the series definitely diverges.

We still need to check what happens at $x = 3$ and $x = -3$.

✳ When $x = 3$, the series becomes $\displaystyle\sum_{n=1}^{\infty} \frac{n3^n}{3^n} = \sum_{n=1}^{\infty} n$. This diverges by the nth term test since $\displaystyle\lim_{n \to \infty} n = \infty \neq 0$. The terms are not getting small.

✳ When $x = -3$, the series becomes $\displaystyle\sum_{n=1}^{\infty} \frac{n(-3)^n}{3^n} = \sum_{n=1}^{\infty} (-1)^n n$. This also diverges by the nth term test, since $\displaystyle\lim_{n \to \infty} (-1)^n n \neq 0$.

This series converges only when $-3 < x < 3$. So we say that the interval of convergence is $(-3, 3)$.

Example 3   Determine the values of x for which the following series converges.

$$\sum_{n=1}^{\infty} n^n x^n$$

Solution   Let's hit it with the root test:

$$\lim_{n \to \infty} \left| n^n x^n \right|^{1/n} = \lim_{n \to \infty} n|x| = \begin{cases} 0 & \text{if } x = 0 \\ \infty & \text{if } x \neq 0 \end{cases}$$

So the series diverges for all x other than 0. At $x = 0$, the series does converge, but that's really not a big deal. When $x = 0$, the series becomes $0 + 0 + 0 + \cdots$ so of course it converges. Every power series will converge at $x = 0$. This particular series illustrates the worst possible case. It diverges everywhere except the one place where it must converge, $x = 0$.

Hey, whatever whirls your dervish.

 ## 4.11 Which test to apply when?

Having a battery of tests to apply does no good if you don't know when to apply them. You don't want a gynecological exam when you have athlete's foot. So in this section, let's talk about the secrets for determining which test to apply. Suppose you have just been handed a series $\sum_{n=1}^{\infty} a_n$. What to do?

Step 1 Apply the nth-term test. If $\lim_{n \to \infty} a_n \neq 0$, the series diverges. Done, just like that. If $\lim_{n \to \infty} a_n = 0$, the series may converge. Proceed to Step 2.

Step 2 Check if the series is a geometric series or a p-series. If it's a geometric series $\sum_{n=0}^{\infty} ar^n$ and $|r| < 1$, then it converges. If it is a p-series $\sum_{n=1}^{\infty} \frac{1}{n^p}$ and $p > 1$, then the series converges. Otherwise, it diverges.

Step 3 If $\sum\limits_{n=1}^{\infty} a_n$ is a positive term series, use one of the following tests.

    1. Basic comparison test

    2. Limit comparison test

       In both of the above tests, you usually compare with either a p-series or a geometric series. In unusual cases, compare the series to a series that can be shown to converge or diverge by some other test, like the integral test.

    3. Ratio test: When there is an $n!$, or a c^n, then there is a good chance this is the test to use.

    4. Root test: When there is an n^n or some function of n to the nth power, this is probably the way to go. Sometimes works with a c^n depending on what else is present.

    5. Integral test: Must have terms a_n that are decreasing toward 0 and that can be integrated.

Step 4 If $\sum\limits_{n=1}^{\infty} a_n$ is an alternating series, use one of the following tests.

    1. Alternating series test

    2. A test from step 3 applied to $\sum\limits_{n=1}^{\infty} |a_n|$. If $\sum\limits_{n=1}^{\infty} |a_n|$ converges, so does $\sum\limits_{n=1}^{\infty} a_n$.

Step 5 For a power series $\sum\limits_{n=1}^{\infty} a_n x^n$, use the ratio or root test to find the interval of convergence. Then use the above tests to check for convergence at the endpoints of the interval.

Let's do some examples of spotting the right test, without working out the details.

Examples

1. $\sum\limits_{n=1}^{\infty} \left(\dfrac{e}{\pi}\right)^n$. It's just a geometric series with $r = e/\pi < 1$, so it converges. For the fun of it, we could use the ratio or root test instead.

2. $\sum_{n=1}^{\infty} \dfrac{(-1)^n}{10^n}$. Taking absolute values gives a geometric series that converges, so this series does also. Or use the alternating series test.

3. $\sum_{n=1}^{\infty} \dfrac{n!}{2^{n^2}}$. The factorial tells us to use the ratio test. It converges.

4. $\sum_{n=1}^{\infty} \dfrac{\sqrt{n}}{n+1}$. The basic comparison test doesn't work. Use the limit comparison test with $\sum_{n=1}^{\infty} \dfrac{1}{\sqrt{n}}$ to see that it diverges.

5. $\sum_{n=2}^{\infty} \dfrac{\sqrt{n}}{n-1}$. Use the basic comparison test with $\sum_{n=2}^{\infty} \dfrac{1}{\sqrt{n}}$ to see that the series diverges.

6. $\sum_{n=1}^{\infty} \dfrac{n!}{(2n)!}$. This definitely calls for the ratio test (look at all those factorials), and the series converges.

Hey, whatever pops your weasel.

 ## 4.12 Taylor series

Polynomials are the type of functions we can get to know intimately. They're open and frank and don't hold back information. Take $f(x) = 2 - 3x + 7x^2$, for instance. This polynomial will tell us anything we want to know and more. We say, "Hey there, f, what's your value at 4?" and $f(x)$ responds, "Well, just take $2 - 3(4) + 7(4^2) = 2 - 12 + 112 = 102$ and there you go. And by the way, I have a weird rash that itches like crazy, but it's getting better."

That's just the kind of functions they are. The amazing part is that just with some additions and multiplications, we can get the value of this polynomial anywhere we want.

The same works for a general polynomial of the form:

$$f(x) = a_0 + a_1 x + a_2 x^2 + \cdots + a_n x^n$$

It can be evaluated at any point using just addition and multiplication, operations we've been familiar with since way before puberty. No secrets, no subterfuge. What you see is what you get.

The same holds true for another form of polynomial:

$$g(x) = b_0 + b_1(x - a) + b_2(x - a)^2 + \cdots + b_n(x - a)^n$$

This is a polynomial in a different pantsuit. If we multiplied out each of the terms $(x - a)^2$, $(x - a)^3$, ..., $(x - a)^n$, and then cleaned up, we would have a polynomial that looked just like the first expression for $f(x)$.

But there are other less friendly functions. Functions that prefer to stay in the basement where no one will try to start up a conversation. Yes, we're talking about $\sin x$, $\ln x$, e^x and the like. And let's not mention $\arctan x$.

Try asking $\ln x$, "So, umm, what's your value at 3?" You'll get a response like, "Why do you want to know? What's it to you? Why're you prying into my affairs? You some kind of hot shot mathhead? Think you can just walk up here and with a little addition and multiplication, you have me figured out? Well, you're out of luck, base 2 brain. It's none of your exponential business what my value at 3 is. Now clear out of here, before I kick you in your tangenital component."

You get the idea, unfriendly And in particular, not about to share with you its value at 3. So what do you do?

Well, that's what this section is all about. We'll replace those nasty functions with our dear friends the polynomials. The hard to compute functions will be approximated by the easy-to-compute polynomials. Pretty clever, huh? Let's start with an extended example.

Example 1   Approximate $f(x) = e^x$ by a polynomial for x near 0.

If we use a fancier high-degree polynomial, the approximation will get better. Let's start simply.

Phase 1   Approximate $f(x) = e^x$ by a polynomial of the form

$$y = a_0 + a_1 x$$

This is a so-called linear polynomial, meaning that its graph is a line. Suppose we want to approximate e^x by a line for values of x near 0. The best choice for such a line is the tangent line to e^x at $(x, y) = (0, 1)$, as in Figure 4.7.

Notice that the tangent line has the same value and the same first derivative as e^x does at $x = 0$. The slope of the tangent line is the derivative of e^x at $x = 0$, since this follows from the definition of the derivative. But $d(e^x)/dx = e^x$, so

$$\left.\frac{d(e^x)}{dx}\right|_{x=0} = e^0 = 1$$

So the slope of the line is 1. Since the line has y-intercept 1, the equation of the line is

$$y = x + 1$$

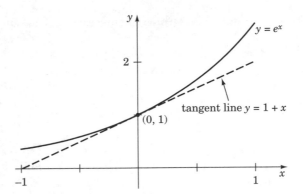

Figure 4.7 Approximating $y = e^x$ by a line.

We have found a line that approximates $y = e^x$ for x near 0, namely, $y = 1 + x$. So

$$e^x \approx 1 + x$$

How accurate is it? Let's evaluate both e^x and $1 + x$ at a small value of x, say $x = 0.1$. Then

$$e^{0.1} \approx 1.105170918$$

On the other hand, $1 + x$ evaluated at $x = 0.1$ is 1.1. So we can see that this "linear approximation" is valid to one decimal place. Not too shabby.

Phase 2 Now we will approximate $f(x) = e^x$ by a second-degree polynomial of the form $g(x) = a_0 + a_1x + a_2x^2$ for values of x near 0. Instead of approximating the graph of e^x by a line, we will be approximating it by a parabola. This should allow us to stay closer to the graph for a longer period of time. This round, we will find the values of a_0, a_1, and a_2 that will ensure that $f(x)$ and the polynomial $g(x)$ have the same value, the same first derivative, and the same second derivative, all at $x = 0$. Then the two functions should be pretty similar, at least for values of x close to $x = 0$.

$$
\begin{aligned}
f(x) &= e^x & &\text{so } f(0) = e^0 = 1 \\
f'(x) &= e^x & &\text{so } f'(0) = e^0 = 1 \\
f''(x) &= e^x & &\text{so } f''(0) = e^0 = 1 \\
g(x) &= a_0 + a_1x + a_2x^2 & &\text{so } g(0) = a_0 \\
g'(x) &= a_1 + 2a_2x & &\text{so } g'(0) = a_1 \\
g''(x) &= 2a_2 & &\text{so } g''(0) = 2a_2
\end{aligned}
$$

For the values and derivatives of these two functions to match at $x = 0$, we need:

$$1 = f(0) = g(0) = a_0 \qquad \text{so } a_0 = 1$$
$$1 = f'(0) = g'(0) = a_1 \qquad \text{so } a_1 = 1$$
$$1 = f''(0) = g''(0) = 2a_2 \qquad \text{so } a_2 = \tfrac{1}{2}$$

therefore, we get that $g(x) = 1 + x + x^2/2$, and so

$$e^x \approx 1 + x + \frac{x^2}{2} \qquad \text{for } x \text{ near } 0$$

This is the same as the first linear approximation that we found, only now we have an additional term involving x^2. The parabolic curve giving us our approximation appears in Figure 4.8.

How good is this approximation? Let's try it when $x = 0.1$. Remember, $e^{0.1} \approx 1.105170918$. Now $1 + x + x^2/2$ at $x = 0.1$ gives 1.105. Not a bad approximation, wouldn't you say? We have picked up another decimal place worth of accuracy.

Phase 3 Let's go one more step and find a third-degree polynomial $g(x) = a_0 + a_1x + a_2x^2 + a_3x^3$ to approximate $f(x) = e^x$ for x near 0.

Again, $f(0) = e^0 = 1, f'(0) = e^0 = 1, f''(0) = e^0 = 1$, and $f'''(0) = e^0 = 1$.

$$g(x) = a_0 + a_1x + a_2x^2 + a_3x^3 \qquad \text{so } g(0) = a_0$$
$$g'(x) = a_1 + 2a_2x + 3a_3x^2 \qquad \text{so } g'(0) = a_1$$
$$g''(x) = 2a_2 + 3(2)a_3x \qquad \text{so } g''(0) = 2a_2$$
$$g'''(x) = 3(2)a_3 \qquad \text{so } g'''(0) = 6a_3$$

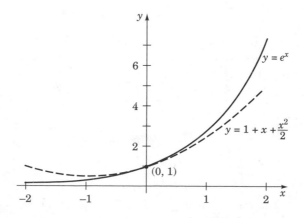

Figure 4.8 Approximating $y = e^x$ by a parabola.

For the values and derivatives of these two functions to match at $x = 0$, we need:

$$1 = f(0) = g(0) = a_0 \qquad \text{so } a_0 = 1$$
$$1 = f'(0) = g'(0) = a_1 \qquad \text{so } a_1 = 1$$
$$1 = f''(0) = g''(0) = 2a_2 \qquad \text{so } a_2 = \tfrac{1}{2}$$
$$1 = f'''(0) = g'''(0) = 3(2)a_3 \qquad \text{so } a_3 = \tfrac{1}{6}$$

This gives

$$e^x \approx 1 + x + \frac{x^2}{2} + \frac{x^3}{6}$$

which is the same as our previous approximation but with an extra $x^3/6$ tacked on the end. The cubic curve that is our new approximation to e^x for x near 0 appears in Figure 4.9.

How accurate is this improved approximation? Remember that $e^{0.1} \approx 1.105170918$. We calculate that $1 + x + x^2/2 + x^3/6$ is $1.1051666\ldots$ when $x = 0.1$. Wow! That's a good approximation. Makes you want to call CNN.

We can see a pattern developing. If we continue to add additional terms in this manner, we would find:

$$e^x \approx 1 + x + \frac{x^2}{2!} + \frac{x^3}{3!} + \frac{x^4}{4!} + \cdots$$

This looks a lot like a power series!

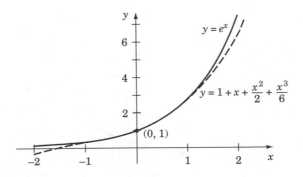

Figure 4.9 Approximating $y = e^x$ by a cubic curve.

Now, you get the idea! We take some ugly difficult function, call it $f(x)$, and we want to find good approximations for $f(x)$ for values of x near some fixed value a. (In our example above, we used $a = 0$.)

So we find a polynomial that behaves very much like $f(x)$ for values of x near a. In general, we will use a polynomial of the form

$$g(x) = b_0 + b_1(x - a) + b_2(x - a)^2 + b_3(x - a)^3 + \cdots + b_n(x - a)^n$$

To make the polynomial behave like $f(x)$ for x near a, we need its value at $x = a$ and the values of its first n-derivatives at $x = a$ to match the same values for $f(x)$.

That is to say, we ask that

$$f(a) = g(a)$$
$$f'(a) = g'(a)$$
$$f''(a) = g''(a)$$
$$f'''(a) = g'''(a)$$
$$\cdots\cdots\cdots\cdots$$
$$f^{(n)}(a) = g^{(n)}(a)$$

Now, straightforward differentiation gives

$$g(x) = b_0 + b_1(x - a) + b_2(x - a)^2 + b_3(x - a)^3 + \cdots + b_n(x - a)^n$$
$$g'(x) = b_1 + 2b_2(x - a) + 3b_3(x - a)^2 + \cdots + nb_n(x - a)^{n-1}$$
$$g''(x) = 2b_2 + 3(2)b_3(x - a) + \cdots + n(n - 1)b_n(x - a)^{n-2}$$
$$g'''(x) = 3(2)b_3 + \cdots + n(n - 1)(n - 2)b_n(x - a)^{n-3}$$
$$\cdots\cdots\cdots\cdots\cdots\cdots\cdots\cdots\cdots\cdots\cdots\cdots\cdots\cdots$$
$$g^{(n)}(x) = n(n - 1)(n - 2)\cdots(3)(2)b_n$$

we must have:

$$f(a) = g(a) = b_0 \qquad \text{so } b_0 = f(a)$$
$$f'(a) = g'(a) = b_1 \qquad \text{so } b_1 = f'(a)$$
$$f''(a) = g''(a) = 2b_2 \qquad \text{so } b_2 = \frac{f''(a)}{2}$$
$$f'''(a) = g'''(a) = 3!b_3 \qquad \text{so } b_3 = \frac{f'''(a)}{3!}$$
$$\cdots\cdots\cdots\cdots\cdots\cdots\cdots\cdots\cdots$$
$$f^{(n)}(a) = g^{(n)}(a) = n!b_n \qquad \text{so } b_n = \frac{f^{(n)}(a)}{n!}$$

Substituting these expressions in for the b's gives us one of the most famous and useful formulas in all of mathematics.

Taylor's Approximation

$$f(x) \approx f(a) + f'(a)(x - a) + \frac{f''(a)}{2!}(x-a)^2 + \frac{f'''(a)}{3!}(x - a)^3 + \cdots + \frac{f^{(n)}(a)}{n!}(x-a)^n$$

This approximation works well for values of x near a. Often, a is just equal to 0. Then the formula simplifies and sometimes gets a new name.

Maclaurin's Approximation

$$f(x) \approx f(0) + f'(0)x + \frac{f''(0)}{2!}x^2 + \frac{f'''(0)}{3!}x^3 + \cdots + \frac{f^{(n)}(0)}{n!}x^n$$

Let's try an example.

Example 2   Find a third-degree Taylor approximation to $f(x) = \sin x$ for $a = 0$, and use it to approximate $\sin(0.05)$.

Solution   Well, $f(x) = \sin x$, $f'(x) = \cos x$, $f''(x) = -\sin x$, and $f'''(x) = -\cos x$.

So $f(0) = \sin 0 = 0$, $f'(0) = \cos 0 = 1$, $f''(0) = -\sin 0 = 0$, and $f'''(0) = -\cos 0 = -1$.

Then

$$\sin x \approx f(0) + f'(0)x + \frac{f''(0)}{2!}x^2 + \frac{f'''(0)}{3!}x^3$$

$$\approx 0 + 1x + 0 + \frac{-1}{3!}x^3$$

$$\approx x - \frac{x^3}{6}$$

Is this a good approximation at $x = 0.05$?

At $x = 0.05$:
$$x - \frac{x^3}{6} = 0.049875$$
$$\sin(0.05) = 0.049979169\ldots$$

Another impressive approximation! Just imagine how good it would have been if we had included a few more terms. In fact, why not do that? It's not hard to see that if we continued to add terms in this example, we would find:

$$\sin x = x - \frac{x^3}{3!} + \frac{x^5}{5!} - \frac{x^7}{7!} + \cdots$$

This is a power series. It is called the *Taylor series* of $\sin x$. When it converges, we can use it to get better and better approximations of our nasty function. But when our power series diverges, it's completely useless.

Hey, whatever tickles your troglodyte.

 4.13 Taylor's formula with remainder

Now we get to the end, the remnant, the stuff that's left over when everything else has been snapped up. Yes, it's time for the remainder sale. Time to move out those Taylor series remainders.

Although Taylor's formula gives us a good approximation to $f(x)$ near $x = a$, it's not exact, and there is an error. And sometimes when we use Taylor's formula we absolutely positively have to know how large the error can be. Otherwise, the bridge we designed almost holds cars. Or the new male contraceptive pill we have created almost prevents pregnancy. Fortunately, there is a simple method for bounding the error.

If we take a degree n Taylor polynomial to approximate $f(x)$ near $x = a$, we get:

Taylor's Formula with Remainder

$$f(x) = \underbrace{f(a) + f'(a)(x - a) + \frac{f''(a)}{2!}(x - a)^2 + \cdots + \frac{f^{(n)}(a)}{n!}(x - a)^n}_{\text{Taylor polynomial } P_n(x)}$$

$$+ \underbrace{\frac{f^{(n+1)}(c)}{(n + 1)!}(x - a)^{n+1}}_{\text{Remainder } R_n(x)}$$

where c is some number between a and x.

The quantity $R_n(x)$ is the *remainder,* or *error term.* It's the difference between the degree n Taylor polynomial at x, which is just an approximation, and the exact value $f(x)$. Notice that it depends on some unidentified number c that lies somewhere between a and x.

Unfortunately, when we use Taylor approximations, and we want to know how big the possible error is, we don't know which c gives the precise error. So the procedure is to bound the value of $R_n(x)$ for any value of c between a and x and therefore get a bound on how bad the error could possibly be.

Example 1 [Estimating sin (1)] Ever wondered about the value of sin (1) where 1 is in radians? (Not to be confused with first sin, which involved eating an apple in a garden.) It's the kind of thing that can keep you up at night. Sure you could plug into your calculator to see what it is, but that's not very satisfying. We would like to be able to see how the calculator actually finds such a value. So let's answer this question so you'll be well rested in the morning. Estimate sin (1) with a third-degree Taylor polynomial and determine a bound for the accuracy of your estimate.

Solution The third-degree Taylor approximation for $\sin x$ with $a = 0$ is $\sin x \approx x - x^3/3!$. This gives an estimate for the value of sin (1) which is

$$\sin (1) \approx 1 - \frac{1^3}{3!}$$
$$\approx 1 - 0.1666\ldots$$
$$= 0.8333\ldots$$

But how accurate is this? The remainder formula tells us that the error $R_3(1)$ is given by

$$R_3(1) = \frac{(\sin x)^{(4)}(c)}{(4)!}(1)^4$$

for some c with $0 \leq c \leq 1$. We don't know what this c is, but amazingly enough, we don't need to know what it is in order to bound $R_3(1)$. Since the fourth derivative $(\sin x)^{(4)}$ is equal to $\sin x$, and the values of $\sin x$ are always between -1 and 1, we know that $|(\sin)^{(4)}(c)| \leq 1$. So

$$|R_3(1)| \leq \frac{1}{4!} = \frac{1}{24}$$
$$\leq 0.042$$

The error is no larger than this number, no matter what the value of c might be. The remainder formula tells us that the true value of sin (1) lies between $(0.8333\ldots - 0.042)$ and $(0.8333\ldots + 0.042)$. Just to verify this, note that the actual value of sin (1) is $0.8414709848079\ldots$, which is within this range.

Example 2 Estimate $\sqrt{e}$ with a second-degree Taylor series and determine a bound for the accuracy of your estimate. Repeat with a tenth-degree Taylor polynomial.

Solution The second-degree Taylor series approximation for e^x about $a = 0$ is $e^x \approx 1 + x + x^2/2!$. Plugging in $x = 0.5$ gives

$$\sqrt{e} = e^{0.5} \approx 1 + (0.5) + \frac{(0.5)^2}{2!} \approx 1 + 0.5 + 0.125 = 1.625$$

So how accurate is this? Well, the remainder formula tells us that the error is

$$R_2(0.5) = \frac{e^v}{3!}(0.5)^3$$

for some c with $0 \leq c < 0.5$. How big can this be? Well, unlike in the previous example, we don't know that e^c is always less than 1. However, e^x is an increasing function, and $e^c \leq e^{0.5}$. Unfortunately, we don't know how big the number $e^{0.5}$ is. In fact that's what we want to find!

But we do know that $e^{0.5} < 4^{0.5} = 2$. So the error $R_2(0.5)$ satisfies

$$|R_2(0.5)| \leq \frac{2}{3!}(0.5)^3 = \frac{1}{24} < 0.042$$

The error is less than 0.042, so we know for sure that $e^{0.5}$ is between $1.625 - 0.042$ and $1.625 + 0.042$. That's not so accurate, but hey, we didn't use that many terms.

If we use the tenth-degree polynomial, $e^x \approx 1 + x + \frac{x^2}{2!} + \cdots + \frac{x^{10}}{10!}$, then plugging in $x = 0.5$ gives us an estimate for $e^{0.5}$ of 1.648721271. The remainder is now bounded by $\frac{2}{11!}(0.5)^{11}$, which is less than 2.45×10^{-11}, a mighty small error for the approximation. So we know just by doing a few multiplications and additions that the true value of $e^{0.5}$ is 1.648721271, accurate to that many decimal places. Hey, we just did as well as our calculator could do. That means we are as smart as a calculator. (Calculators fall somewhere between toasters and coffeemakers on the appliance intelligence scale.)

And speaking of remainders, did you hear about the two student entrepeneurs from the Math Club who bet their year's tuition that red T-shirts covered with yellow calculus equations would be the big fashion hit on campus that spring? They put all the the funds their parents had sent together with all they could borrow to purchase a mountain of T-shirts. Unfortunately, the end of winter classes came without any orders. Despite a massive ad campaign in the college paper, the phone failed to ring even once. Unsold T-shirts (remainders) filled their dorm room. Without money for tuition, ruin was imminent. Then, a week before fees were due, the phone rang. "This is Bud from the college bookstore" says the voice at the other end. "Hey, word has it that you have red T-shirts with some yellow equations on them. We could use them for our spring promotion. We can pay top dollar." Bud mentioned a massive sum of money, but with one caveat. "I just have to clear it with purchasing because it's such a big amount. If you don't hear from me by 5 p.m. Friday, it's a done deal."

The two students couldn't believe their luck. They sat around nervously as the clock moved toward 5 p.m. Friday, just praying they wouldn't get a call. At 4:55 p.m. the phone rang. Nervously, her heart beating furiously, one of the students picked up the phone. Then, slowly, as she listened, a smile crossed her face. "Good news," she said. "Your car is on fire."

4.14 Some famous Taylor series

We have already mentioned a few of the most famous Taylor series. For instance,

$$e^x = 1 + x + \frac{x^2}{2!} + \frac{x^3}{3!} + \cdots$$

and

$$\sin x = x - \frac{x^3}{3!} + \frac{x^5}{5!} - \cdots$$

In fact, we can manipulate Taylor series just like functions in order to generate other series. For instance, since $\cos x$ can be obtained by differentiating $\sin x$, a Taylor series for $\cos x$ can be obtained by differentiating the Taylor series for $\sin x$ term by term. The result is

$$\cos x = 1 - \frac{x^2}{2!} + \frac{x^4}{4!} - \cdots$$

What other basic Taylor series do we have? Let's go back to geometric series for just a second. Remember that when $|r| < 1$, we knew that the geometric series

$$a + ar + ar^2 + \cdots$$

converged to $\dfrac{a}{1-r}$. Letting $a = 1$ and replacing r with the variable x, we have

$$\boxed{\dfrac{1}{1-x} = 1 + x + x^2 + \cdots \qquad \text{for any } |x| < 1}$$

So this is a series representation of $\dfrac{1}{1-x}$. In fact, it agrees with what we would obtain by applying Taylor's formula. Notice though that it is only valid for $|x| < 1$.

> **Warning**   The Taylor series of a function may diverge for some or all values of x.

The four basic series above are worth memorizing. They can often be used to derive other series when the need arises.

Example   Find a series that gives $\arctan x$ for $|x| < 1$.

Solution   You may recall from when we studied antiderivatives that

$$\arctan x = \int_0^x \dfrac{1}{1+t^2}\, dt$$

Or you may not. But anyway this was one of the formulas we worked out then.

Now, replacing each x in $\dfrac{1}{1-x} = 1 + x + x^2 + \cdots$ by $-t^2$, we obtain

$$\dfrac{1}{1+t^2} = 1 - t^2 + t^4 - \cdots$$

Integrating both sides gives us

$$\arctan x = \int_0^x \dfrac{1}{1+t^2}\, dt = \int_0^x 1 - t^2 + t^4 - \cdots \, dt$$

$$= t\Big|_0^x - \dfrac{t^3}{3}\Big|_0^x + \dfrac{t^5}{5}\Big|_0^x - \cdots = x - \dfrac{x^3}{3} + \dfrac{x^5}{5} - \cdots$$

which is the Taylor series for $\arctan x$ for $|x| < 1$. But this series is only valid when $|x| < 1$ even though $\arctan x$ is defined for all real numbers.

Hey, whatever buttons your cardigan.

Vectors: From Euclid to Cupid

In this chapter we introduce vectors, in the plane and in three-dimensional space. Vectors are like arrows, and have both a direction and a length. They come up when we think about things like force or velocity, which have both a direction in which they point or push and a magnitude or size corresponding to a length.

5.1 Vectors in the plane

We won't spend a lot of time in the plane, as befits three-dimensional creatures like ourselves. But we will get flat for a little while. It will help to clarify our 3D world when we return to it. Let's start with distance.

Very Basic Fact The distance between two points in the xy-plane, $P_1 = (x_1, y_1)$ and $P_2 = (x_2, y_2)$, is given by

$$d(P_1, P_2) = \sqrt{(x_2 - x_1)^2 + (y_2 - y_1)^2}$$

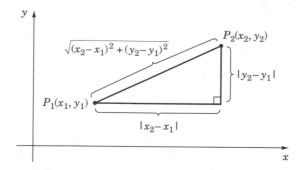

Figure 5.1 The Pythagorean theorem tells us how to measure distance between points in the plane.

This rule comes from Pythagoras, that guy who wore Toxas and a sheet. Put P_1 and P_2 down in the plane and form a right triangle as in Figure 5.1. One edge has length $|x_2 - x_1|$ and the other edge has length $|y_2 - y_1|$, so the hypotenuse has length $\sqrt{(x_2 - x_1)^2 + (y_2 - y_1)^2}$.

Now it's time to talk about vectors. Remember Camp Whatsitoya, where your parents sent you for 3 months that one summer, supposedly to experience the wonders of nature, but really because they just couldn't stand to be around you for another second? And you know how the wonders of nature consisted primarily of mosquitos the size of Piper aircraft, and woods with a couple of scraggly pines growing in a field of poison ivy?

And that entire summer, as you scratched yourself raw, the one thing that you looked forward to was archery. There was nothing quite like the zing of that arrow cutting through the air, heading straight and true for the target across the field. And as you watched it sink home, you would imagine it was your brother Everett, who was also at the camp, and who was the one person on Earth who could make you suffer more than the poison ivy. No, we haven't gone off on this tangent (yes, that's a little math pun, get used to it) because we have early signs of age-related dementia. We're reminding you of all this because that arrow represents a vector. Think of that arrow as having been lofted into space, and then freeze time, so that all motion stops. Two quantities are relevant for that arrow as it sits frozen in space. The first is its length. After all, a long arrow can be shot a lot farther than a short one. The second relevant quantity is the direction in which the arrow is headed. What could be more important than that? The direction in which it heads will determine whether it ends up dead center in the bullseye or dead center in the rear of one of the campers waiting in line down at the latrine.

We are not concerned at this point with where the arrow sits in space. We usually think of it situated with its tail at the origin, without changing

Figure 5.2 A vector is determined by its length and its direction. These are all the same vector.

its direction or its length. Then we use its other endpoint, the sharp end, to describe the vector. So the vector in Figure 5.2 is given by $\mathbf{v} = \langle 2, 1 \rangle$. We call the numbers 2 and 1 the *components* of the vector. It can change its position, and yet be the same vector. It doesn't matter if the feather end is at the origin or at the point $(321, 467)$. It's still the same arrow, it's just gone for a walk.

We might as well define a scalar while we're at it. A *scalar* is a real number. Why a new name for real number? Just to emphasize the difference between vectors and scalars. Scalars are real numbers and vectors are not. Alternately, scalars just have a magnitude, but vectors have both a magnitude and a direction.

The vector beginning at $(2, 1)$ and ending at $(1, 3)$ in Figure 5.3, when translated over so its tail sits at the origin, gives a vector that begins at $(0, 0)$ and ends at $(-1, 2)$.

If a vector begins at (x_1, y_1) and ends at (x_2, y_2), it's given by

$$\mathbf{v} = \langle x_2 - x_1, y_2 - y_1 \rangle$$

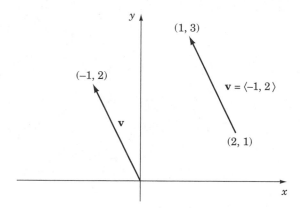

Figure 5.3 An arrow starting at (2, 1) gives the same vector as one starting at the origin.

The length of a vector is the distance from its initial point P to its final point Q. The length of **v** is written $|\mathbf{v}|$. When we place a vector so that it begins at the origin, its length is given by the Pythagorean theorem.

Pythagorean Theorem   If $\mathbf{v} = \langle x, y \rangle$, then $|\mathbf{v}| = \sqrt{x^2 + y^2}$.

The length of a vector $|\mathbf{v}|$ is also called the *norm* or the *magnitude* of the vector.

OPERATIONS ON VECTORS

Even though they aren't numbers, we can still add and subtract vectors. This is done componentwise. If $\mathbf{v} = \langle x_1, y_1 \rangle$ and $\mathbf{w} = \langle x_2, y_2 \rangle$, then

$$\mathbf{v} + \mathbf{w} = \langle x_1 + x_2, y_1 + y_2 \rangle$$

and

$$\mathbf{v} - \mathbf{w} = \langle x_1 - x_2, y_1 - y_2 \rangle$$

Addition and subtraction have geometric interpretations as well. Adding two vectors is the same as putting the first vector with its tail at the origin, and placing the second vector so that its tail begins at the tip of the first vector. The sum vector begins at the origin and ends at the tip of the second vector. Note that $\mathbf{v} + \mathbf{w} = \mathbf{w} + \mathbf{v}$, since we create a parallelogram with the same diagonal when adding them either way.

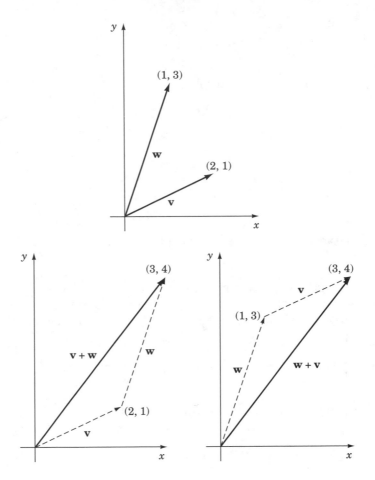

Figure 5.4 Adding two vectors. It doesn't matter in which order we add them.

Group Project Form groups of two. Lie down in the plane (also called the floor). Imagine that each of you is a vector and take your vector sum (see Figure 5.4). Consider how humiliated the second vector feels, having the first vector point to its tail. And empathize with the first vector, having to point to the second vector's tail. Discuss which vector is more oppressed.

For subtracting a vector, reverse the direction on the second vector before placing it at the end of the first vector, as in Figure 5.5. The order does matter when subtracting.

We can multiply a vector by a real number (or scalar) c. This is the same as stretching it by a factor of c. The formula for multiplying $\mathbf{a} = \langle a_1, a_2 \rangle$

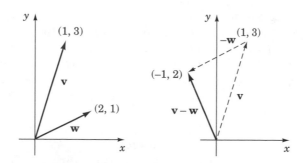

Figure 5.5 Subtracting two vectors. Order matters.

by c is

$$c\mathbf{a} = \langle cu_1, cu_2 \rangle$$

This is called *scalar multiplication*. Scalar multiplication by a positive number does NOT change the direction of the vector, it just changes the length. Multiplication by a negative number flips the vector to the opposite direction, and multiplication by 0 makes a vector lose all direction.

RULES VECTORS LIVE BY

1. $\mathbf{a} + \mathbf{b} = \mathbf{b} + \mathbf{a}$ (Alice and Bill in a limo is the same as Bill and Alice in a limo.)

2. $\mathbf{a} + (\mathbf{b} + \mathbf{c}) = (\mathbf{a} + \mathbf{b}) + \mathbf{c}$ (Alice joining Bill and Craig, who are already in the limo, is the same as Alice and Bill joining Craig in the limo.)

3. $c(\mathbf{a} + \mathbf{b}) = c\mathbf{a} + c\mathbf{b}$ (Taking c copies of Alice and Bill in the limo is the same as taking c copies of Alice and c copies of Bill in the limo.)

4. $(c + e)\mathbf{a} = c\mathbf{a} + e\mathbf{a}$ (If we make $c + e$ clones of Alice, and stick them in the limo, that is the same as making c clones and then e clones, and sticking them all in the limo.)

5. $(ce)\mathbf{a} = c(e\mathbf{a}) = e(c\mathbf{a})$ (If we make ce clones for the limo, that's the same as c clones of e clones, etc. This is starting to become a stretch.)

A VERY SPECIAL VECTOR

Every vector has a length and a direction. But in fact, there is an exception, called the *zero vector,* given by $\mathbf{0} = \langle 0, 0 \rangle$. Of course, most of us don't think much of zero, the number. There's not a lot to say about it. It's the big Nada.

Hence the expression, when talking about someone of whom you don't have a high opinion, "What a zero." But the zero vector is even lower than the number zero. Not only does it lack any length, it doesn't have any direction either. So someone who is adrift in life, uncertain about a direction to take, living at home with his or her parents at the age of 36, flipping burgers for pocket change and having to borrow the minivan to go out on weekends might be called a zero vector.

[*Editor's note*: If you are that person, please do not sue this publishing company for defamation. Sue the authors directly. Contact us for their addresses.]

5.2 Space: The final (exam) frontier

Okay, let's talk turkey. All of single-variable calculus (you know, the calculus we have done up to now) takes place in the xy-plane. That is to say, it occurs in a two-dimensional world. Pancake flat, unvarying, level as level can be. So flat, it makes Nebraska look mountainous. It's so completely and totally flat that the grooves in a CD look like the Himalayas in comparison. We are talking falattt, smooth, constant, no change, no nothing, next gas station 2000 miles, not a bump, crumb, or dip for as far as the eye can see. Get the idea? Horifreaking-zontal. But do we live in a flat world like that? Of course not. We live in the full bloom of three dimensions, where everything has a solid, robust roundness to it. In our world, nothing is really completely two-dimensional. Even a playing card or a piece of Saran Wrap has some thickness. So it's time to notch up our belts and take the math to another level. We want to describe the world we live in, that beautiful, juicy, sensuous, nectar-filled three-dimensional world we fondly call home. From now on, points will have three coordinates instead of two. We're talking the full (x, y, z), from here on out. That's right. We are now in three-dimensional space, fondly called *3-space,* and going by the formal name of R^3 (pronounced "R three").

DISTANCE BETWEEN TWO POINTS

Let's work out the formula for the distance between two points $P(x_0, y_0, z_0)$ and $Q(x_1, y_1, z_1)$ in 3-space. Ducks can do this. When flying south for the winter, a duck easily spots the hunters in the duck blind. Then it turns to the duck next to it and says, "Remington T20 double gauge shotgun, South Southwest, at 350 meters as the duck flies. Range of that model is 200 meters, so we're clear. Pass it down." The V-shaped formations that ducks fly in makes it easy to pass information up and down the line. The downfall of the ducks is their short attention span. They get distracted by some other duck that gets a little out of formation, and all the ducks start laughing, really quacking it up. By the time they calm down, it's too late. BLAM, fewer mallards going south this winter.

So the ducks can do all of the calculations, and then some, but they're weak on the follow-through. You, on the other hand, have the potential to maintain your concentration, meaning you may get to celebrate your A on the midterm at Fort Lauderdale over spring break.

Here is the formula.

Distance Formula   If d is the distance between points $P(x_1, y_1, z_1)$ and $Q(x_2, y_2, z_2)$, then

$$d = \sqrt{(x_2 - x_1)^2 + (y_2 - y_1)^2 + (z_2 - z_1)^2}$$

Where did this formula come from? It's just the Pythagorean theorem applied twice, as in Figure 5.6. Notice that $|x_2 - x_1|$ is the length of the box in the x-direction. And $|y_2 - y_1|$ is the length of the box in the y-direction. To find the length of the diagonal δ across the bottom, we use the fact it's the hypotenuse of a right triangle. So

$$\delta = \sqrt{(x_2 - x_1)^2 + (y_2 - y_1)^2}$$

The distance from P to Q is the diagonal of the box, which is the hypotenuse of a right triangle with base δ and height $|z_2 - z_1|$. So the Pythagorean theorem gives

$$d(P(x_1, y_1, z_1), Q(x_2, y_2, z_2)) = \sqrt{\left(\sqrt{(x_2 - x_1)^2 + (y_2 - y_1)^2}\right)^2 + (z_2 - z_1)^2}$$
$$= \sqrt{(x_2 - x_1)^2 + (y_2 - y_1)^2 + (z_2 - z_1)^2}$$

Hey, whatever pops your bubble wrap.

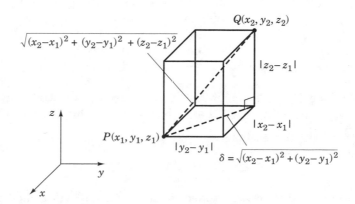

Figure 5.6 The Pythagorean theorem lets us measure distance between points in 3-space.

SPHERES IN SPACE

We all know about spheres, especially the famous one we all live on. That's one big round ball, the Earth (or for you alien readers, Mars).

The way to pin down a particular sphere in space is to give its center $P(a, b, c)$, and its radius r. Then if $Q(x, y, z)$ is any old point on the sphere,

$$d(P(a, b, c), Q(x, y, z)) = \sqrt{(x - a)^2 + (y - b)^2 + (z - c)^2} = r$$

Squaring both sides we get this general equation.

Equation for a sphere of radius r centered at $P(a, b, c)$

$$(x - a)^2 + (y - b)^2 + (z - c)^2 = r^2$$

Example $(x - 3)^2 + (y - 4)^2 + (z + 1)^2 = 16$ is the equation of a sphere of radius 4 centered at the point $P(3, 4, -1)$.

Hey, whatever steams your clams.

5.3 Vectors in space

As we mentioned, a vector should be thought of as an arrow. But now we'll allow the arrow to fly in three-dimensional space, rather than in the plane.

As before, two things determine a vector, its direction and its length. Usually we place the vector so it begins at the origin, and then it is given by the coordinates of its other endpoint. So the vector $\mathbf{v} = \langle 2, 3, 4 \rangle$ is drawn to start at the point $P(0, 0, 0)$ and end at the point $Q(2, 3, 4)$, as in Figure 5.7.

The length of a vector is the distance from its initial point P to its final point Q. If we place it so that it begins at the origin, and if it is given by $\mathbf{v} = \langle x, y, z \rangle$, then

$$|\mathbf{v}| = \sqrt{x^2 + y^2 + z^2}$$

Vectors in 3-space are added, subtracted, and multiplied by scalars just like vectors in the plane, but with an extra component trailing along. The zero vector is even more of a zero than the one in the plane, $\mathbf{0} = \langle 0, 0, 0 \rangle$.

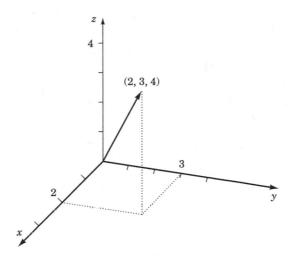

Figure 5.7 The vector $\mathbf{v} = \langle 2, 3, 4 \rangle$.

UNIT VECTORS

A vector that has length 1 is called a *unit* vector. "One what?" you might ask. Well, one unit! It's one unit long, hence the name. There are a whole wagonload of unit vectors, one for each possible direction.

Handy Fact   We can start with any vector $\mathbf{v}$ (except the zero vector) and stretch or shrink it to get a unit vector. Starting with $\mathbf{v}$,

$$\mathbf{u} = \frac{\mathbf{v}}{|\mathbf{v}|}$$

is a unit vector pointing in the same direction as $\mathbf{v}$.

We can check that the length of $\mathbf{u}$ really is 1. How?

$$|\mathbf{u}| = \left| \frac{\mathbf{v}}{|\mathbf{v}|} \right| = \frac{|\mathbf{v}|}{|\mathbf{v}|} = 1$$

Here are a few more members of the Vector Hall of Fame:

$$\mathbf{i} = \langle 1, 0, 0 \rangle$$
$$\mathbf{j} = \langle 0, 1, 0 \rangle$$
$$\mathbf{k} = \langle 0, 0, 1 \rangle$$

These are also known as Izzy, Jane, and Klondike. All three are exactly one unit long, and they point along the positive x-, y-, and z-axes, respectively.

Why are they called $\mathbf{i}$, $\mathbf{j}$, and $\mathbf{k}$? Some say that at the math conference where they were named, the best candidate was "I Just don't Know." Another theory is that all the other good letters were already taken: a, b, c for constants, d for derivative, e for $2.71828\ldots$, f and g for functions, and h for Preparation H. You quickly find yourself staring at $\mathbf{i}$, $\mathbf{j}$, and $\mathbf{k}$.

We can combine multiples of $\mathbf{i}$, $\mathbf{j}$, and $\mathbf{k}$ to describe any vector. Let $\mathbf{a} = \langle a_1, a_2, a_3 \rangle$ be a vector. Then we can rewrite it as

$$\mathbf{a} = \langle a_1, a_2, a_3 \rangle = \langle a_1, 0, 0 \rangle + \langle 0, a_2, 0 \rangle + \langle 0, 0, a_3 \rangle$$
$$= a_1 \langle 1, 0, 0 \rangle + a_2 \langle 0, 1, 0 \rangle + a_3 \langle 0, 0, 1 \rangle$$
$$= a_1 \mathbf{i} + a_2 \mathbf{j} + a_3 \mathbf{k}$$

So we see

$$\boxed{\langle a_1, a_2, a_3 \rangle = a_1 \mathbf{i} + a_2 \mathbf{j} + a_3 \mathbf{k}}$$

Mathematicians tend to go with the bracket notation, or with parentheses, like (a_1, a_2, a_3), but physicists consider a vector undressed unless it sports an $\mathbf{i}$, $\mathbf{j}$, and $\mathbf{k}$.

Hey, whatever tightens your wingnut.

5.4 The dot product

The dot product takes two vectors, puts them together, and produces a real number. This is different from two vectors falling in love, getting married, and having a baby vector. We will come to that shortly. The baby in this case is not another vector. It's a real number. So this is more like two vectors falling in love, getting married, and giving birth to a porpoise. How do you do that? (Take the dot product of two vectors, not give birth to a porpoise.)

Given vectors $\mathbf{a} = \langle a_1, a_2, a_3 \rangle$ and $\mathbf{b} = \langle b_1, b_2, b_3 \rangle$, their dot product is the number:

$$\boxed{\mathbf{a} \cdot \mathbf{b} = a_1 b_1 + a_2 b_2 + a_3 b_3}$$

Example 1 (Calculating a Dot Product) Let $\mathbf{a} = \langle 3, 2, 1 \rangle$ and $\mathbf{b} = \langle 2, -1, 4 \rangle$. Then $\mathbf{a} \cdot \mathbf{b} = 3(2) + 2(-1) + 1(4) = 8$.

Most people call it the *dot product*, but once in a while you will hear someone call it the *inner product* (don't ask).

Often, people ask plaintively, "Exactly what is a dot product?" The dot product is essentially a compatibility index for the two vectors. It measures how similar two vectors are. If the two vectors point in approximately the same direction, we get a positive dot product, and the more lined up they are, the larger it is (unless the vectors are very short). If the two vectors are perpendicular, the dot product is zero; if the two vectors point in approximately opposite directions, the dot product is negative.

THE DOT PRODUCT AS A MATCHMAKING TOOL

We can use the dot product to match people up with their perfect mate, be it animal, mineral, or vegetable.

Here's how. Each person fills out a form stating his or her preferences for the following three categories, rated from -5 to 5, where -5 expresses extreme dislike and 5 expresses adoration.

1. Body piercing

2. Sushi

3. Classical music

For each person, we create a vector in 3-space with his or her answers to questions 1, 2, and 3 as the three components in the vector. This is called the *compatibility vector*. Then, to determine the compatibility of two people, we take the dot product of their compatibility vectors. The larger the result, the more compatible they are.

For example, Evelyn Wankel decides to look for love. She goes into Calcthrob, the new singles matching service, and fills out her preferences for these categories, resulting in vector $\mathbf{E} = \langle 0, 5, 4 \rangle$. Meanwhile, in another part of the building, Lucien and Joe Bob are filling out their preferences, yielding vectors $\mathbf{L} = \langle -5, -2, 5 \rangle$ and $\mathbf{J} = \langle 5, 5, 5 \rangle$. When Evelyn turns her card in at the desk, the attendant, Dr. Lace, Psychology Ph.D., says, "Please wait here, while I run this through our computer."

She goes in the back and asks, "Hey, Ed, got any?" Without looking up from his book, Ed hands her the two cards containing $\mathbf{L}$ and $\mathbf{J}$. Dr. Lace takes the dot product of each with $\mathbf{E}$, finding that $\mathbf{E} \cdot \mathbf{L} = 10$ and $\mathbf{E} \cdot \mathbf{J} = 45$.

"Bring in $\mathbf{J}$," she says to Ed, as she returns to where Evelyn waits. "I am thrilled to say we found a match. You're going to find this hard to believe, but by coincidence, he's right here in the building. I've asked Mr. Wisteria to bring him over. Ah! Here they are now."

Ed shows in a large grizzled man wearing a gray sleeveless T-shirt that covers neither his belly nor the large dagger tattooed onto his arm. "Hey, there. Name's Joe Bob. You are looking real good, *real* good," he says, as he eyes her up and down.

"Oh," said Evelyn, "you're not exactly what I was expecting."

"The computer knows you better than you know yourself," says Dr. Lace.

As they leave arm in arm, Joe Bob says, "So you like classical music, too? I got some REO Speedwagon in the pick-up."

RULES DOT PRODUCTS LIVE BY

1. $\mathbf{a} \cdot \mathbf{a} = |\mathbf{a}|^2$ (This says that you are compatible with yourself. Your dot product with yourself is always positive and equals your magnitude squared. This is lucky. If you weren't compatible with yourself, you'd be living with someone you don't like inside your own head, a recipe for disaster. There is one exception. The **0**-vector is incompatible with everyone, including itself.)

2. $\mathbf{a} \cdot \mathbf{b} = \mathbf{b} \cdot \mathbf{a}$ (If **a** is compatible with **b**, then **b** is compatible with **a**.)

3. $c\mathbf{a} \cdot \mathbf{b} = \mathbf{a} \cdot c\mathbf{b} = c(\mathbf{a} \cdot \mathbf{b})$ (Scalars can float through dot products like they own the place.)

GEOMETRIC INTERPRETATION OF THE DOT PRODUCT

The dot product means something geometric, too. Take any two nonzero vectors **a** and **b**, placed so they begin at the same point. Let θ be the angle between them; it can be any angle with $0 \le \theta \le \pi$. Then

$$\mathbf{a} \cdot \mathbf{b} = |\mathbf{a}||\mathbf{b}| \cos \theta$$

This is easy to show if you know the law of cosines. Don't worry if you don't; no one's gone to jail for breaking this law. Worst that's ever happened is a small fine and some community service. This has an interesting consequence. Suppose that **a** and **b** are any two nonzero vectors. Then

The vectors **a** and **b** are perpendicular if and only if $\mathbf{a} \cdot \mathbf{b} = 0$.

Why? From our rule above, $\mathbf{a} \cdot \mathbf{b} = 0$ exactly when $\cos \theta = 0$. But this happens if and only if $\theta = \pi/2$. Now we have a very easy test to apply to decide whether or not two vectors are perpendicular. Let's try it.

Example 2   The Whitlessberg Town Council decides to build a new street crossing Main Street to accommodate the proliferation of knickknack stores. If Main Street begins at $(0, 0)$ and ends at $(3, 4)$ (coordinates are in miles with the Town Hall at the center, x-axis east and y-axis north) and the new street crosses Main and runs from $(2, 5)$ to $(4, 3)$, determine if the two streets cross perpendicularly.

Solution The first street has the vector $\mathbf{v} = \langle 3, 4 \rangle$ pointing in its direction. The second street has vector $\mathbf{w} = \langle 4 - 2, 3 - 5 \rangle = \langle 2, -2 \rangle$ pointing in its direction. Since the dot product $\mathbf{v} \cdot \mathbf{w} = 3(2) + 4(-2) = -2$ is not zero, the sad truth is that the two streets do not meet at right angles.

Just as poets have 20 different words to express yearning, mathematicians have lots of words for their favorite concepts. Right angles are way up there in the pantheon of hot ideas for math types. So we have "perpendicular," "orthogonal," and "normal," all of which mean the same thing. The word "orthogonal" comes from the Greek root *ortho,* which means upright or right-angled or standard, as in "Orthodontists make your teeth right" or "Orthoweed killer makes your lawn right." Be careful not to confuse these last two examples. We also have orthodoxy, the belief in the standard system, and even orthography, the theory of correct spelling. (You have to wonder about the competing theory.)

We can also use this formula to calculate angles between vectors.

The angle θ between vectors $\mathbf{a}$ and $\mathbf{b}$ is determined by

$$\cos \theta = \frac{\mathbf{a} \cdot \mathbf{b}}{|\mathbf{a}||\mathbf{b}|}$$

Example 3 An airport runway points in the direction given by the vector $\mathbf{a} = \langle 2, 1, 0 \rangle$ and a plane is coming in with velocity vector $\mathbf{b} = \langle 3, 2, -1 \rangle$. Find the angle between the runway and the approach vector of the airplane.

Solution We know

$$\cos \theta = \frac{\mathbf{a} \cdot \mathbf{b}}{|\mathbf{a}||\mathbf{b}|} = \frac{8}{\sqrt{5}\sqrt{14}} \approx 0.9562$$

So, $\theta \approx \arccos (0.9562) \approx 0.2971$ radian $\approx 17°$.

Time to bank the plane.

PROJECTION OF ONE VECTOR ONTO ANOTHER

In working with vectors, we often need to find the perpendicular shadow of one vector onto another, called the projection. (See Figure 5.8.)

We denote the length of this shadow of $\mathbf{a}$ on $\mathbf{b}$ by $\text{comp}_b \, \mathbf{a}$ ($\text{comp}_b \, \mathbf{a}$ stands for "component in the direction of $\mathbf{b}$ of $\mathbf{a}$"). Notice that the length of

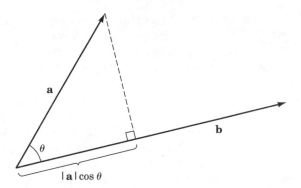

Figure 5.8 Projecting vector **a** onto vector **b** gives a shadow of length comp$_b$ **a**.

the shadow of **a** on **b** is given by $|\mathbf{a}|\cos\theta$. So, from the boxed formula above, we have

$$\text{comp}_b\,\mathbf{a} = |\mathbf{a}|\cos\theta = |\mathbf{a}|\left(\frac{\mathbf{a}\cdot\mathbf{b}}{|\mathbf{a}||\mathbf{b}|}\right) = \frac{\mathbf{a}\cdot\mathbf{b}}{|\mathbf{b}|}$$

If we want to find the vector that is the projection of **a** on **b**, denoted proj$_b$ **a**, we can just multiply comp$_b$ **a** by the unit vector in the direction of **b**.

$$\text{proj}_b\,\mathbf{a} = (\text{comp}_b\,\mathbf{a})\frac{\mathbf{b}}{|\mathbf{b}|}$$

$$= \left(\frac{\mathbf{a}\cdot\mathbf{b}}{|\mathbf{b}|}\right)\frac{\mathbf{b}}{|\mathbf{b}|}$$

$$= \frac{\mathbf{a}\cdot\mathbf{b}}{|\mathbf{b}|^2}\mathbf{b}$$

Example 4   Find the projection of the vector $\mathbf{a} = \langle 1,\,3,\,2\rangle$ to the vector $\mathbf{b} = \langle -1, 2, 4\rangle$.

Solution   This is easy.

$$\text{proj}_b\,\mathbf{a} = \frac{\mathbf{a}\cdot\mathbf{b}}{|\mathbf{b}|^2}\mathbf{b} = \frac{-1+6+8}{(-1)^2+(2)^2+(4)^2}\langle -1, 2, 4\rangle = \frac{13}{21}\langle -1, 2, 4\rangle$$

WORK

Physics has a very specific definition for "work" which may not be what you think. Suppose you spend the entire afternoon cleaning the apartment; you get down on your knees and scrub the bathroom floor, clean in those hard to reach places, make your roommate's bed, the works. You lift the piano off the rug to clean underneath it. Finally you put everything neatly back where it started out. Your roomie comes home at the end of the day, and you say, "See all the work I've done?" But if she's a physics major, she says, "Work, what work? You haven't done any work. To do work, you must drag an object across the floor and then compute the force applied to the object times the distance that the object was pulled. You haven't done any work at all." This is an example of why it is not a good idea to have firearms in the house.

In physics, the work done by a constant force in moving an object along a straight line is given by the dot product of the force vector $\mathbf{F}$ with the displacement vector $\mathbf{D}$,

$$\boxed{\text{Work } W = \mathbf{F} \cdot \mathbf{D}}$$

The displacement vector $\mathbf{D}$ is the vector that begins where the object started and ends where the object ended. If the force vector and the displacement vector line up, the work is just the size of the force vector times the distance moved (which is the size of the displacement vector). On the other hand, if the force vector and the displacement vector are at right angles, the dot product will be 0, and the force does no work when the object is moved.

The units used for force are called *newtons* in the metric system, and the unit used for work is the *joule*. One joule is the work done when an object is moved 1 meter against a force of 1 newton. The word "joule" comes from Newton's rival Leibniz who said, "A jewel to whoever moves against Newton."

[*Editor's note:* The authors made this up. I apologize.]

Example 5   Fifi drags her owner along a sidewalk that's 200 meters long. If Fifi (she's a poodle, although I guess that's obvious from the name) exerts a force of 2 newtons on the leash, and the leash is at an angle of 45° from the ground, how much work does Fifi do?

Solution   As in Figure 5.9, we can assume that Fifi and her owner start at (0, 0) in the xy-plane and move 200 meters along the x-axis. So the displacement vector is given by $\mathbf{D} = \langle 200, 0 \rangle$. The force vector has length 2 and has angle 45° from the x-axis. It therefore points in the direction of the

Figure 5.9 The force of Fifi.

vector $\langle 1, -1 \rangle$. This is not a vector of length 2, however. A unit vector pointing along $\langle 1, -1 \rangle$ is obtained by dividing $\langle 1, -1 \rangle$ by its length $\sqrt{2}$, giving the vector $\langle \sqrt{2}/2, -\sqrt{2}/2 \rangle$. The force vector **F** therefore is twice this, and is given by

$$\mathbf{F} = \langle \sqrt{2}, -\sqrt{2} \rangle$$

The inner product of the displacement vector $\langle 200, 0 \rangle$ with the force vector $\langle \sqrt{2}, -\sqrt{2} \rangle$ gives the work done, which is therefore equal to $200\sqrt{2}$ joules.

If the force vector isn't constant, we need to break the displacement into lots of teeny, nearly straight segments, along which the force vector is close to constant. We then add up the work done in all these little steps. The limit of this process gives an integral, which computes the work done by a nonconstant force along a nonstraight displacement. Later we'll calculate quantities like this using line integrals.

5.5 The cross product

DETERMINANTS

Perhaps Arnold Schwarzenegger's greatest unknown movie was "The Determinator." Never screened, the movie showed Arnold sitting at a desk with a pencil and paper, working out determinants. It goes on like this for an hour and a half. Occasionally he turns to the camera and says, "Hasta la vista, baby." Finally he gets frustrated, pulls out a submachine gun and a few grenades and gets into a gunfight with the Cleveland State Math Department. That's the highlight of the movie.

So just what were those determinants he was working on? Given a 2×2 array of four numbers, their determinant is

$$\det \begin{pmatrix} a & b \\ c & d \end{pmatrix} = \begin{vmatrix} a & b \\ c & d \end{vmatrix} = ad - bc$$

Instead of writing "det," we can use vertical bars around the array to represent the determinant. So

$$\begin{vmatrix} 2 & 1 \\ 3 & -1 \end{vmatrix} = 2(-1) - 1(3) = -5$$

Given a 3×3 array of numbers, we can take its determinant as follows:

$$\det \begin{pmatrix} a & b & c \\ d & e & f \\ g & h & j \end{pmatrix} = \begin{vmatrix} a & b & c \\ d & e & f \\ g & h & j \end{vmatrix} = a \begin{vmatrix} e & f \\ h & j \end{vmatrix} - b \begin{vmatrix} d & f \\ g & j \end{vmatrix} + c \begin{vmatrix} d & e \\ g & h \end{vmatrix}$$

Example (3×3 Determinant)

$$\begin{vmatrix} 1 & 2 & 4 \\ -1 & 3 & 0 \\ 2 & 5 & -2 \end{vmatrix} = 1 \begin{vmatrix} 3 & 0 \\ 5 & -2 \end{vmatrix} - 2 \begin{vmatrix} -1 & 0 \\ 2 & -2 \end{vmatrix} + 4 \begin{vmatrix} -1 & 3 \\ 2 & 5 \end{vmatrix}$$
$$= 1(-6) - 2(2) + 4(-11) = -54$$

Determinants are very important in an area of mathematics called linear algebra. Linear algebra is essential for a variety of applications, from economics to computer graphics. While we don't have to know too much about determinants to understand calculus, we do need to know how to calculate them for 2×2 and 3×3 matrices.

DETERMINANTS AND THE CROSS PRODUCT

Now we are going to take two vectors, tell them to multiply, and the result will be a new vector. Kind of like putting two people in a room, waiting nine months, and then out come three people. Of course, this doesn't work so well if the two people are of the same gender. So we can think of the cross product as the result of two vectors meeting, falling in love, and having a child, who then grows up very tall and very very skinny, as vectors do. (See Figure 5.10.)

The formula for the cross product of vectors $\mathbf{a} = \langle a_1, a_2, a_3 \rangle$ and $\mathbf{b} = \langle b_1, b_2, b_3 \rangle$ is

$$\mathbf{a} \times \mathbf{b} = \langle a_2 b_3 - a_3 b_2, a_3 b_1 - a_1 b_3, a_1 b_2 - a_2 b_1 \rangle$$

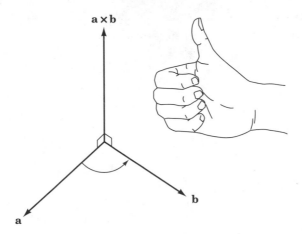

Figure 5.10 The cross product $\mathbf{a} \times \mathbf{b}$ is perpendicular to each of $\mathbf{a}$ and $\mathbf{b}$. Its direction is given by the thumb of the right hand when your fingers rotate from $\mathbf{a}$ to $\mathbf{b}$.

Remembering this formula would be a mental feat comparable to swimming the English Channel while reciting the "Bhagvat Gita." Luckily, no one has to remember it. There is an easy way to get it.

$$\mathbf{a} \times \mathbf{b} = \begin{vmatrix} \mathbf{i} & \mathbf{j} & \mathbf{k} \\ a_1 & a_2 & a_3 \\ b_1 & b_2 & b_3 \end{vmatrix}$$

Taking the determinant of the matrix with first row $\mathbf{i}, \mathbf{j}, \mathbf{k}$, second row $\mathbf{a}$, and third row $\mathbf{b}$ gives $\mathbf{a} \times \mathbf{b}$.

We can check this by taking the determinant:

$$\begin{vmatrix} \mathbf{i} & \mathbf{j} & \mathbf{k} \\ a_1 & a_2 & a_3 \\ b_1 & b_2 & b_3 \end{vmatrix} = \mathbf{i}(a_2 b_3 - a_3 b_2) - \mathbf{j}(a_1 b_3 - a_3 b_1) + \mathbf{k}(a_1 b_2 - a_2 b_1)$$

$$= \langle a_2 b_3 - a_3 b_2, \, a_3 b_1 - a_1 b_3, \, a_1 b_2 - a_2 b_1 \rangle$$

Would you look at that? A perfect match with our previous formula for the cross product.

Example 2   Find $\mathbf{a} \times \mathbf{b}$ when $\mathbf{a} = \langle 1, 2, -1 \rangle$ and $\mathbf{b} = \langle -2, 3, 1 \rangle$.

Solution

$$\begin{vmatrix} \mathbf{i} & \mathbf{j} & \mathbf{k} \\ 1 & 2 & -1 \\ -2 & 3 & 1 \end{vmatrix} = \mathbf{i}\begin{vmatrix} 2 & -1 \\ 3 & 1 \end{vmatrix} - \mathbf{j}\begin{vmatrix} 1 & -1 \\ -2 & 1 \end{vmatrix} + \mathbf{k}\begin{vmatrix} 1 & 2 \\ -2 & 3 \end{vmatrix}$$

$$= \mathbf{i}(5) - \mathbf{j}(-1) + \mathbf{k}(7) = \langle 5, 1, 7 \rangle$$

As is often the case in the human world, it turns out that the child vector is not compatible with the parents. The kid is a rebel, completely incompatible, perpendicular to both.

> **Important Fact** $\mathbf{a} \times \mathbf{b}$ is perpendicular to both $\mathbf{a}$ and $\mathbf{b}$.

How to see this? To show that two vectors are perpendicular, we just need to show that their dot product is 0. Let's see that $\mathbf{a}$ is perpendicular to $\mathbf{a} \times \mathbf{b}$:

$$\mathbf{a} \cdot (\mathbf{a} \times \mathbf{b}) = \langle a_1, a_2, a_3 \rangle \cdot \langle a_2 b_3 - a_3 b_2, a_3 b_1 - a_1 b_3, a_1 b_2 - a_2 b_1 \rangle$$

$$= a_1(a_2 b_3 - a_3 b_2) + a_2(a_3 b_1 - a_1 b_3) + a_3(a_1 b_2 - a_2 b_1)$$

$$= 0$$

All the terms cancel out, and we get a dot product equal to 0. So these two vectors really are perpendicular. The same holds for the dot product of $\mathbf{b}$ and $\mathbf{a} \times \mathbf{b}$.

Keep in mind that you can only take the cross product of two vectors and that the result will be a vector. Numerous jokes have been created to help you remember this fact. We'll throw in a couple of new ones, as well.

CROSS PRODUCT JOKES

Question: What do you get when you cross a lion with a tiger?
Answer: A vector perpendicular to both.

Question: What do you get when you cross a lion with a mountain climber?
Answer: Nothing, you can't cross a vector with a scalar.[*]

Question: What do you get when you cross a mosquito with a fish monger?
Answer: Nothing, you can't cross a (disease) vector with a (fish) scaler.

---

[*]"Lion" is supposed to sound like "line," which makes you think of "vector," and "mountain climber"—well, you figure it out. (Hey, don't look at us. We didn't make this one up.)

Question: What do you get when you cross a mathematician with a movie star?
Answer: Dream on, it'll never happen.

Question: Why did the chicken cross the road?
Answer: To get a vector perpendicular to itself and the road.

Question: If you take the vector **A** and cross it with the vector **Dresser**, what do you get?
Answer: A cross dresser.

Okay, okay, so they aren't funny. But math jokes seldom are. Usually, the pleasure comes from relishing their truly pathetic nature.

GEOMETRY OF THE CROSS PRODUCT

We have talked about the direction of the cross product vector $\mathbf{a} \times \mathbf{b}$. It is perpendicular to each of $\mathbf{a}$ and $\mathbf{b}$, and it obeys the right-hand rule. But what about its magnitude?

The magnitude of $\mathbf{a} \times \mathbf{b}$ is given by

$$\boxed{|\mathbf{a} \times \mathbf{b}| = |\mathbf{a}||\mathbf{b}|\sin\theta}$$

Here again, θ is the angle between the vectors $\mathbf{a}$ and $\mathbf{b}$. Notice how similar it is to the formula we saw earlier: $\mathbf{a} \cdot \mathbf{b} = |\mathbf{a}||\mathbf{b}|\cos\theta$. Be careful not to confuse these two.

Also notice that the magnitude of the cross product does not depend on the order in which we take the cross product. In other words,

$$|\mathbf{a} \times \mathbf{b}| = |\mathbf{b} \times \mathbf{a}|$$

By the right-hand rule, $\mathbf{a} \times \mathbf{b}$ will point in exactly the opposite direction from $\mathbf{b} \times \mathbf{a}$. Two vectors that point in opposite directions and that have the same magnitude are just the negative of one another. In other words,

$$\boxed{\mathbf{a} \times \mathbf{b} = -\mathbf{b} \times \mathbf{a}}$$

That's a fact worth remembering.

YES, BUT WHAT'S THE CROSS PRODUCT GOOD FOR?

How many times have you said to that special someone in your life, "Honey, what's a simple formula for finding the area of a triangle when you have two vectors that give two of the adjacent edges?"

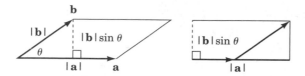

Figure 5.11 Cross products let us measure areas of parallelograms and triangles.

And that special someone responded, "I need to find a new special someone in my life."

Well, you needn't worry that this scene will be repeated anymore, as there is a simple formula that we will tell you about now.

Let's look at the parallelogram in Figure 5.11 whose two adjacent sides are **a** and **b**. We can cut a right triangle off the left-hand side of the parallelogram and shift it over to the right side to create a rectangle with the same height and base length as the parallelogram. So

$$\text{Area (parallelogram)} = (\text{base})(\text{height}) = |\mathbf{a}||\mathbf{b}|\sin\theta = |\mathbf{a} \times \mathbf{b}|$$

A triangle with two adjacent sides **a** and **b** has area half that of the parallelogram, so we get the formula:

$$\boxed{\text{Area (triangle)} = \tfrac{1}{2}|\mathbf{a}||\mathbf{b}|\sin\theta = \tfrac{1}{2}|\mathbf{a} \times \mathbf{b}|}$$

This brings us to our next joke:

Question: What do you get when you cross a hippo with a giraffe?
Answer: A vector perpendicular to each whose length is the length of the hippo times the length of the giraffe times the sine of the angle between them.

Hey, whatever blackens your catfish.

THE SCALAR TRIPLE PRODUCT, OR THAT'S SOME PARALLELEPIPED YOU'RE PACKING

Now, we can get really weird and take the product of THREE vectors. This isn't really new; it's just a combination of cross and dot products. If we do this just right, the number we get gives the volume of the parallelepiped spanned by the three vectors. This is sort of like the skewed box that you'd get if you start with a carton and step on it until the three edges coming out of a

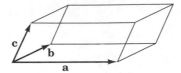

Figure 5.12 Triple products give the volume of a parallelepiped.

corner give the three vectors shown in Figure 5.12. Incidentally, being able to pronounce "parallelepiped" is a sure-fire way to impress anyone.

The formula for the triple product of vectors $\mathbf{a}$, $\mathbf{b}$, $\mathbf{c}$ is $\mathbf{a} \cdot (\mathbf{b} \times \mathbf{c})$. If you get confused and instead calculate $(\mathbf{a} \times \mathbf{b}) \cdot \mathbf{c}$, don't worry—you'll get the same answer. The answer is a number, that is to say a scalar, and its absolute value is the volume of the parallelepiped we discussed.

Notice that we must do the cross product first and then the dot product. It makes no sense to try to calculate $(\mathbf{a} \cdot \mathbf{b}) \times \mathbf{c}$. Doing the dot product first gives a scalar, and we can't take the cross product of a scalar with a vector. And notice that after taking the dot product and cross product, the result is a number; the outer vertical bars mean to take the absolute values, so we can drop any minus sign at the end.

Instead of actually taking the cross product and then the dot product, there is a simple formula for finding the scalar triple product:

$$\mathbf{a} \cdot (\mathbf{b} \times \mathbf{c}) = \begin{vmatrix} a_1 & a_2 & a_3 \\ b_1 & b_2 & b_3 \\ c_1 & c_2 & c_3 \end{vmatrix}$$

One application of this idea is to determine if three vectors lie in a plane. Three vectors lie in a plane precisely when the volume of the parallelepiped they span is zero.

Example   Do the vectors $\mathbf{a} = \langle 1, 2, 3 \rangle$, $\mathbf{b} = \langle 3, 2, 1 \rangle$, and $\mathbf{c} = \langle 8, 4, 0 \rangle$ lie in a common plane?

Solution   Taking the scalar triple product gives

$$\mathbf{a} \cdot (\mathbf{b} \times \mathbf{c}) = \begin{vmatrix} 1 & 2 & 3 \\ 3 & 2 & 1 \\ 8 & 4 & 0 \end{vmatrix} = 0$$

The volume of the parallelepiped they span is zero and we conclude that the three vectors do lie in a common plane.

More applications of the ideas from this chapter will appear in subsequent sections. As the chicken says, "We'll cross that vector when we come to it."

Hey, whatever chokes your throttle.

5.6 Lines in space

Remember how you learned to graph lines in the xy-plane? You would have an equation like $y = 2x + 3$, and you would determine all of the points (x, y) that satisfy that equation. In particular, in this case, the slope is 2 and the y-intercept is 3, so it's easy to describe the line.

Life (and lines) won't be quite that simple in 3-space. Here, instead of one equation, we will use three equations to describe a line. Let's start by figuring out what we need to determine a line. Suppose that we have a point P on the line and a vector $\mathbf{v}$ in the direction of the line. Then there is a unique line that passes through the point in the direction of the vector. The point and the vector completely determine the line.

This is just the way life is, too. One point and a direction often determine a life. For instance, Christopher Columbus (see Figure 5.13) went down in

Spain

USA

Figure 5.13 Christopher Columbus's lifeline was determined by a starting point and a direction.

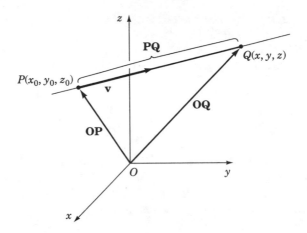

Figure 5.14 A line is determined by a point P and a vector $\mathbf{v}$.

history at the point when Queen Isabella provided three boats and he headed for the Americas. One point and a direction determined his "lifeline." We should all be so lucky.

Suppose we have one point $P(x_0, y_0, z_0)$ that is on the line and a vector $\mathbf{v} = \langle a, b, c \rangle$ in the direction of the line. Take $Q(x, y, z)$ to be any other point on the line. Then the vector $\mathbf{PQ}$ points along the same line as the vector $\mathbf{v}$. This means that vector $\mathbf{PQ}$ is a scalar multiple of $\mathbf{v}$. So there is a scalar t with $\mathbf{PQ} = t\mathbf{v}$.

Notice also in Figure 5.14 that we can sum the vectors $\mathbf{OP}$ and $\mathbf{PQ}$ to get $\mathbf{OQ}$. So we have

$$\mathbf{OQ} = \mathbf{OP} + \mathbf{PQ}$$
$$\mathbf{OQ} = \mathbf{OP} + t\mathbf{v}$$

Writing the vectors out, we have

$$\langle x, y, z \rangle = \langle x_0, y_0, z_0 \rangle + t\langle a, b, c \rangle$$

Component by component, we get

$$x = x_0 + ta$$
$$y = y_0 + tb$$
$$z = z_0 + tc$$

These are the *parametric equations* of the line through the point $P(x_0, y_0, z_0)$ with direction vector $\mathbf{v} = \langle a, b, c \rangle$.

The variable t is called the *parameter*. As t varies, the point moves along the line. It's like a bug crawling along the line. When $t = 0$, the bug is at the initial point $P(x_0, y_0, z_0)$. When $t = 1$, the bug is at the end of the vector **v** in Figure 5.15. As t continues to increase the bug will continue to crawl farther out along the line. It's not much of a life, but bugs don't actually expect much anyway. If your brain pan's as big as the head of a needle, you don't have vast expectations.

The same line can have tons of different parametric equations describing it. We can take any point on the line for P and any vector parallel to the line for **v**.

We can also write the equations of a line a little differently. We can solve each of the three equations in a parametrized line for t and then set the t-values equal. This gives the *symmetric equations* for a line:

$$\frac{x - x_0}{a} = \frac{y - y_0}{b} = \frac{z - z_0}{c}$$

Example   A bee makes a beeline through the points $P_0(1, 0, -1)$ and $P_1(2, 3, -2)$. Find parametric and symmetric equations for the line flown by the bee.

Solution   To spice up this example, two points are used to determine the line instead of a point and a direction vector. To write down the parametric equations, we need a point on the line and a vector in the direction of the line. Hey, a point on the line we've got—we have two of them. To get a vector in the line, just take the vector from one to the other, from P_0 to P_1. This

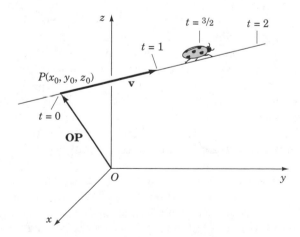

Figure 5.15 The parameter t determines a point on the line.

is $\mathbf{v} = \mathbf{P_0P_1} = \langle 2 - 1, 3 - 0, -2 - (-1) \rangle = \langle 1, 3, -1 \rangle$. The components of this vector are our a, b, and c. We'll use the coordinates of P_0 for x_0, y_0, and z_0. So our parametric equations become

$$x = 1 + t$$
$$y = 0 + 3t$$
$$z = -1 + (-1)t$$

If we switch the roles of P_0 and P_1, we get the parametric equations

$$x = 2 + -t$$
$$y = 3 + -3t$$
$$z = -2 + t$$

These (and many others) are perfectly good parametric equations for the line.

The symmetric equations are obtained by solving each of the parametric equations for t, and setting them all equal. So, using the first set of parametric equations, we have

$$\frac{x - 1}{1} = \frac{y - 0}{3} = \frac{z + 1}{-1}$$

5.7 Planes in space

The word "plane" comes from the Latin word *planus,* which means flat, or level, no features to speak of. Hence, the Latin expression "Planus Janus," which we now say as "Plain Jane" or "Planus in the Anus," which means uncomfortable chair. We would like some means of keeping track of planes in space.

Suppose we are on a flying carpet and we want it to take us to the Snack Bar so we can impress everyone there. Unlike what occurs in cartoons, a flying carpet can neither speak nor understand English. So how do we get it to go where we want it to go? Well, the truth of the matter is that there is a joystick sticking right out of the bathmat, perpendicular to it. Since the carpet must stay perpendicular to the joystick, we can steer it. In other words, this perpendicular stick determines the orientation of the rug in space. (To get the carpet to go forward or backward you give it a good kick.)

The same idea can be used to describe a plane in space. Let's take a vector $\mathbf{n} = \langle a, b, c \rangle$ perpendicular to the plane, which we call a *normal vector.* (There are no abnormal vectors. "Normal" is just another word for "perpendicular." On the other hand, there are abnormal Victors. Take for instance Victor Lamazz of 34 West Huntington Blvd., Chicago.) Then we take a point $P_0 = (x_0, y_0, z_0)$

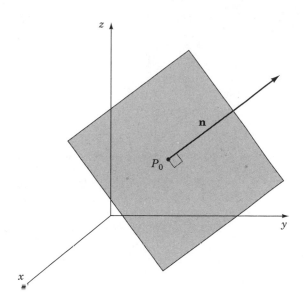

Figure 5.16 A plane is determined by a point P_0 and a normal vector **n**.

which lies on the plane. If P_0 is sitting or standing on the plane, that is also acceptable. The normal vector and the point completely determine the plane. (See Figure 5.16.)

THE EQUATION FOR A PLANE

Given a normal vector $\mathbf{n} = \langle a, b, c \rangle$ and a point $P_0 = (x_0, y_0, z_0)$ on the plane, let's find the equation for the plane. Suppose that $Q(x, y, z)$ is any point at all on the plane.

As in Figure 5.17, the vector $\mathbf{P_0Q}$ lies in the plane. So it is perpendicular to **n** and therefore

$$\mathbf{n} \cdot \mathbf{P_0Q} = 0$$
$$\langle a, b, c \rangle \cdot \langle x - x_0, y - y_0, z - z_0 \rangle = 0$$

So the points (x, y, z) on the plane satisfy the equation

$$\boxed{a(x - x_0) + b(y - y_0) + c(z - z_0) = 0}$$

Multiplying out and collecting terms, this can be rewritten as

$$\boxed{ax + by + cz = d}$$

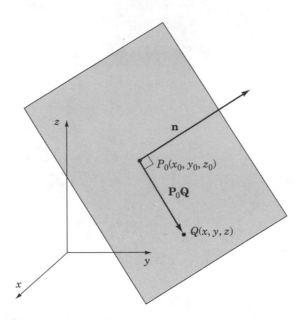

Figure 5.17 The vector $\mathbf{P_0Q}$ is perpendicular to $\mathbf{n}$.

where $d = a(x_0) + b(y_0) + c(z_0)$ is just the collection of all of the constant terms. These are two forms for the equation of a plane in space.

Example 1 (The Tilted Floor)　To save money on a building project, you hired your cousin as a contractor. Now you find that the floor doesn't seem quite level. When you stand at the point $P(1, 2, 3)$ (in the kitchen), you measure a normal vector $\mathbf{n} = \langle 4, 5, 3 \rangle$. Find the equation of the plane containing your floor.

Solution　Plugging the information into the equation for a plane, we have

$$4(x - 1) + 5(y - 2) + 3(z - 3) = 0$$

This simplifies to

$$4x + 5y + 3z = 23$$

Example 2　Your cousin says you must have measured the normal vector incorrectly. He takes you out and together you measure that the corners of

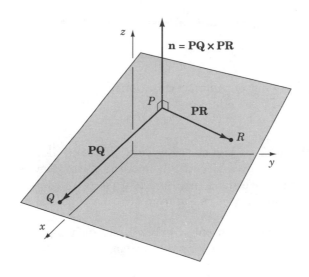

Figure 5.18 The normal vector as the cross product of **PQ** and **PR**.

your triangular living room are at $P(1, 2, 3)$, $Q(5, 0, 1)$, and $R(0, 4, 1)$. Use this to recalculate the equation of the plane containing your floor.

Solution We have to find the normal vector. Since all three points lie in the plane, the vectors **PQ** and **PR** lie in the plane. Their cross product is perpendicular to the plane and will make a good normal vector, as in Figure 5.18.

$$\mathbf{PQ} = \langle 5 - 1, 0 - 2, 1 - 3 \rangle = \langle 4, -2, -2 \rangle$$
$$\mathbf{PR} = \langle 0 - 1, 4 - 2, 1 - 3 \rangle = \langle -1, 2, -2 \rangle$$
$$\mathbf{PQ} \times \mathbf{PR} = \langle 8, 10, 6 \rangle$$

Then the equation for the plane is

$$a(x - x_0) + b(y - y_0) + c(z - z_0) = 0$$

where $\langle a, b, c \rangle = \langle 8, 10, 6 \rangle$ and we can take $(x_0, y_0, z_0) = (1, 2, 3)$. This yields

$$(8)(x - 1) + (10)(y - 2) + (6)(z - 3) = 0$$

After dividing through by 2, this gives the same equation we derived before:

$$4x + 5y + 3z = 23$$

Not too level. Oh well, life is full of hard lessons. Next time use your brother-in-law.

Example 3   Find parametric equations for the line of intersection of the two planes given by the equations $x + y - z = 7$ and $2x - 3y + z = 3$.

Solution   This is a classic type of problem, like the 1964 MGB roadster is to the automotive circuit. To find the parametric equations for the line of intersection, we need two things, a point on the line and a vector in the direction of the line, as in Figure 5.19.

Step 1 First let's find a point on the line. Such a point $P(x, y, z)$ lies in both planes and therefore satisfies both equations. So we are looking for a simultaneous solution to

$$x + \ y - z = 7$$
$$2x - 3y + z = 3$$

This is a set of two equations in three unknowns, so we expect lots of solutions. Not surprising since the two planes intersect in a line, and so

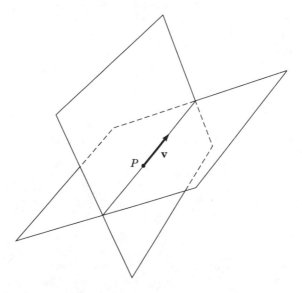

Figure 5.19 The intersection of two planes is a line.

there's a whole line of solutions. We can cut down to one solution by taking $x = 0$. The equations become

$$y - z = 7$$
$$-3y + z = 3$$

Adding them, we have

$$-2y = 10$$
$$y = -5$$

From the first equation, we have $-5 - z = 7$ and $z = -12$. So a point on the line is $P(0, -5, -12)$. Many other points would have worked just as well.

Step 2 Find a vector in the direction of the line.

We can read off normal vectors for the two planes from the coefficients in their equations. A normal vector for the first plane is $\mathbf{n}_1 = \langle 1, 1, -1 \rangle$ and a normal vector for the second plane is $\mathbf{n}_2 = \langle 2, -3, 1 \rangle$. Now for the cool part. Since these vectors are perpendicular to their respective planes, if we find a vector perpendicular to both, it must lie in both planes. That means it lies in the intersection of the two planes, which is our line. How to find a vector perpendicular to both? Hey, that would be the cross product. So we take

$$\mathbf{n} = \mathbf{n}_1 \times \mathbf{n}_2$$

and we have

$$\mathbf{n}_1 \times \mathbf{n}_2 = \begin{vmatrix} \mathbf{i} & \mathbf{j} & \mathbf{k} \\ 1 & 1 & -1 \\ 2 & -3 & 1 \end{vmatrix} = -2\mathbf{i} - 3\mathbf{j} - 5\mathbf{k}$$

So the parametric equations for the line of intersection are

$$\begin{aligned} x &= 0 + t(-2) &&= -2t \\ y &= -5 + t(-3) &&= -5 - 3t \\ z &= -12 + t(-5) &&= -12 - 5t \end{aligned}$$

If instead of the direction vector $\langle -2, -3, -5 \rangle$, we take the opposite vector $\langle 2, 3, 5 \rangle$, we get another option for parametric equations of the same line, namely,

$$\begin{aligned} x &= 2t \\ y &= -5 + 3t \\ z &= -12 + 5t \end{aligned}$$

Hey, whatever sets your Jell-O.

6

Parametric Curves in Space: Riding the Roller Coaster

In this chapter, we'll look at fancy versions of functions which produce vectors instead of numbers. These will be useful for a variety of purposes, particularly for describing curves in space.

Our previous look at functions resembled the study of a lazy pet dog. We fed it a number and it drooled out an answer (another number) onto the carpet. But now we'll look at a function with values in space. This is more like a cobra. We feed it a number, but it doesn't just dribble out a number for an answer. It shoots a stream of venom straight toward our eyes—in other words, it produces a vector.

6.1 Parametric curves

One of the many famous musical pieces about parametric curves is "The Flight of the Bumblebee," composed by Nicolai Rimsky-Korsakov. After listening to the piece, it is clear that Rimsky-Korsakov must have derived precise mathematical equations describing a bee's flight as it traveled from flower to flower collecting pollen, and then set the equations to music.

[*Editor's note:* There has been no independent verification of this theory. Most likely, they made this up, too.]

You, too, can compose beautiful parametric curves. In fact, we have already done this when we described lines. For example, suppose we have a line given by the parametric equations

$$x = 2 + t$$
$$y = -1 + 3t$$
$$z = 4 - 2t$$

We can express all three equations in a single *vector-valued function* of t given by

$$\mathbf{r}(t) = \langle 2 + t, \ -1 + 3t, 4 - 2t \rangle$$

The equations for x, y, and z become the x-, y-, and z-components of the vector. As t varies, $\mathbf{r}(t)$ gives different points along the line, as in Figure 6.1.

We can use this idea of a vector-valued function to model curves more complicated than lines. We call them *parametric curves*. For instance, we could look at the following parametric curve.

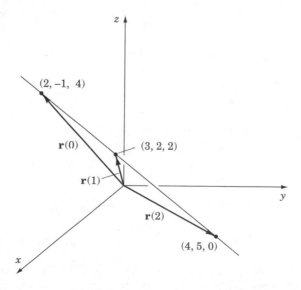

Figure 6.1 The line given by $\mathbf{r}(t) = \langle 2 + t, \ -1 + 3t, 4 - 2t \rangle$.

Example 1

$$\mathbf{r}(t) = \langle \sin t, \ \cos t, t \rangle$$

or, in other words,

$$\mathbf{r}(t) = \sin t\mathbf{i} + \cos t\mathbf{j} + t\mathbf{k}$$

This is like the flight path of an inebriated bumblebee who knows which flowers contain the fermented nectar and treats his job as one long bar hop. The function $\mathbf{r}(t)$ tells us where the bee is after t seconds. So at time $t = 0$, when we're just getting started, the bee is located at $\mathbf{r}(0) = \mathbf{j}$ or at $\langle 0, \ 1, \ 0 \rangle$. At time $t = 20$, after 20 seconds, the bee has buzzed over to $\mathbf{r}(20) = \sin(20)\mathbf{i} + \cos(20)\mathbf{j} + 20\mathbf{k}$, or $\langle \sin(20), \ \cos(20), \ 20 \rangle$.

Drawing a picture of a parametric curve can sometimes be tricky, but in this case it's not so bad. As t increases, the first two coordinates wind around a circle, $\langle \sin t, \ \cos t \rangle$, while the third coordinate increases steadily. This leads to a curve called a helix, which plays a starring role in your genes, where it models the shape of your DNA. (See Figure 6.2.)

More generally, any vector-valued function with three components can be written in the form $\mathbf{r}(t) = \langle x(t), y(t), z(t) \rangle$. In most cases, computers give the best ways of drawing these curves. This being calculus, we're going to differentiate these babies.

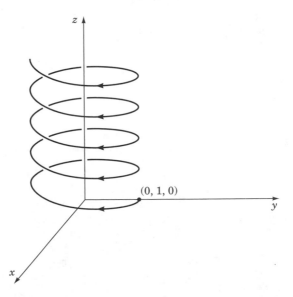

Figure 6.2 A parametrized curve called the helix.

DIFFERENTIATING VECTOR-VALUED FUNCTIONS

The way to do this couldn't be easier. We just differentiate each of the components $x(t)$, $y(t)$, and $z(t)$.

$$\mathbf{r}'(t) = \langle \mathbf{x}'(t), \mathbf{y}'(t), \mathbf{z}'(t) \rangle$$

The vector $\mathbf{r}'(t)$ is called the *velocity vector* to the curve at time t, or sometimes the *tangent vector,* since if we draw it starting at the point on the curve where we computed it, it is tangent to the curve and pointing in the direction of motion. (See Figure 6.3.)

Notice that the derivative is a vector, not a number! To see how our position changes as we move along a curve, we need a direction AND a number that tells us how fast we are moving in that direction. In other words, we need a vector. The number $|\mathbf{r}'(t)|$, the magnitude of the velocity vector, tells us how fast we are moving in the direction of the vector. It is called the *speed.*

The velocity vector of the helix is $\mathbf{r}'(t) = \langle \cos t, \ -\sin t, 1 \rangle$. At time $t = 0$ the tangent vector is $\langle 1, 0, 1 \rangle$. This means that at the instant $t = 0$, the x- and z-coordinates are increasing, but the y-coordinate is not changing since its derivative is 0.

The speed is given by taking the length

$$|\mathbf{r}'(t)| = \sqrt{\cos^2 t + (-\sin t)^2 + 1^2} = \sqrt{2}$$

for any t. So we are moving at a constant speed of $\sqrt{2}$, even as we twist and turn along the helix.

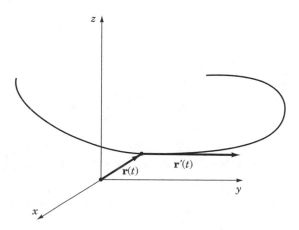

Figure 6.3 The velocity vector $\mathbf{r}'(t)$ is tangent to the curve at $\mathbf{r}(t)$.

RULES TO DIFFERENTIATE CURVES BY

There are a variety of rules for differentiating dot products, cross products, etc., of vector-valued functions. They work just like the product, chain, and sum rules for functions. There are several types of multiplication that can be done: scalar multiplication, dot products, cross products. The appropriate version of the product rule works for all of them. For understanding these rules, keep in mind that $\mathbf{r}(t)$ and $\mathbf{s}(t)$ are vector-valued functions of t, where t is just a real variable like x or y. The function $f(t)$ is a plain old real-valued function of t, or a scalar function as it's sometimes called in vector calculus. Last, and maybe least, is c, which is just a constant real number.

1. $\dfrac{d}{dt}\left[\mathbf{r}(t) + \mathbf{s}(t)\right] = \dfrac{d}{dt}\,\mathbf{r}(t) + \dfrac{d}{dt}\,\mathbf{s}(t)$

2. $\dfrac{d}{dt}\left[c\mathbf{r}(t)\right] = c\dfrac{d}{dt}\,\mathbf{r}(t)$

3. $\dfrac{d}{dt}\left[f(t)\mathbf{r}(t)\right] = f(t)\mathbf{r}'(t) + f'(t)\mathbf{r}(t)$

4. $\dfrac{d}{dt}\left[\mathbf{r}(t) \cdot \mathbf{s}(t)\right] = \mathbf{r}(t) \cdot \mathbf{s}'(t) + \mathbf{r}'(t) \cdot \mathbf{s}(t)$

5. $\dfrac{d}{dt}\left[\mathbf{r}(t) \times \mathbf{s}(t)\right] = \mathbf{r}(t) \times \mathbf{s}'(t) + \mathbf{r}'(t) \times \mathbf{s}(t)$

6. $\dfrac{d}{dt}\left[\mathbf{r}(f(t))\right] = \mathbf{r}'(f(t))f'(t)$

The last one is the *chain rule* for vector-valued functions.

INTEGRATING VECTOR-VALUED FUNCTIONS

Integration of a vector-valued function $\mathbf{r}(t) = \langle x(t), y(t), z(t)\rangle$ is also done by term.

$$\int_a^b \mathbf{r}(t)\,dt = \left\langle \int_a^b x(t)\,dt,\ \int_a^b y(t)\,dt,\ \int_a^b z(t)\,dt \right\rangle$$

When the integral is not a definite integral, a constant is tacked onto the end, just like in regular integration, only this time the constant is a vector:

$$\int \mathbf{r}(t)\,dt = \left\langle \int x(t)\,dt,\ \int y(t)\,dt,\ \int z(t)\,dt \right\rangle + \mathbf{C}$$

Example 2   A mosquito starts at time $t = 0$ at the point $(1, 1, 1)$. From there it flies along a curve for 1 second, after which it lands for lunch. Its velocity vector at time t is given by $\mathbf{v}(t) = \langle t, t^2, t^3 \rangle$. Your ear is located at $(3/2, 4/3, 5/4)$ and your roommate's nose is at $(1/2, 1/3, 1/4)$. Which of you gets bitten?

Solution The solution is obtained by integrating $\mathbf{v}(t)$ to find the position vector $\mathbf{r}(t)$.

$$\mathbf{r}(t) = \int \mathbf{v}(t)\, dt = \left\langle \int t\, dt, \int t^2\, dt, \int t^3\, dt \right\rangle$$

$$= \left\langle \frac{t^2}{2}, \frac{t^3}{3}, \frac{t^4}{4} \right\rangle + \mathbf{C}$$

We still need to figure out what the constant vector $\mathbf{C}$ equals. We can use the fact that we know the position vector at time $t = 0$.

$$\langle 1, 1, 1 \rangle = \mathbf{r}(0) = \langle 0, 0, 0 \rangle + \mathbf{C}$$

So $\mathbf{C} = \langle 1, 1, 1 \rangle$ and

$$\mathbf{r}(t) = \left\langle \frac{t^2}{2} + 1, \frac{t^3}{3} + 1, \frac{t^4}{4} + \right\rangle$$

Where's the little bloodsucker at time $t = 1$?

$$\mathbf{r}(1) = \langle 3/2, 4/3, 5/4 \rangle$$

Looks like your ear is on the mosquito menu after all.

LENGTHS OF CURVES IN SPACE

Suppose you're flying a plane. Your global positioning system tells you where you are at any time. You do some loop de loops, some wing salutes, some stomach-turning maneuvers. As you gracefully land the plane, your copilot looks up from the airsickness bag and asks, "So just how far did we go anyway?"

You now have a typical arc-length problem. What is the length of the curve you traced out as you flew your plane around?

To understand this question, we first have to realize how little we knew about measuring lengths before we learned calculus. The only things whose lengths we really understood were straight lines and maybe a circle or two. We learned how to measure them way back in elementary school. Now we're going to extend our knowledge to the length of wiggly curves.

We do this by approximating the curve by little straight segments and adding up their lengths. This approximation becomes better if we take the segments shorter and use more of them. It becomes exact if we take a limit. Hey, when we take a limit of sums of things which break an interval into more and shorter segments—that's an integral.

So suppose we have a curve $\mathbf{r}(t) = \langle x(t), y(t), z(t) \rangle$ and we want to measure its arc length S from the point corresponding to $t = a$ to the point corresponding to $t = b$. A little bit of the curve from $\mathbf{r}(t)$ to $\mathbf{r}(t + \Delta t)$ has length which we call Δs. For very small Δt, we can assume that we aren't twisting around much over this time interval and we are moving in essentially a straight line. But when we move in a straight line, the distance traveled is the speed times the time interval. So, as in Figure 6.4,

$$\Delta s \approx |\mathbf{r}'(t)|\,\Delta t = \sqrt{(x'(t))^2 + (y'(t))^2 + (z'(t))^2}\,\Delta t$$

Adding up all of these little lengths gives us an approximation to the arc length of the curve:

$$S \approx \sum \Delta s = \sum \sqrt{(x'(t))^2 + (y'(t))^2 + (z'(t))^2}\,\Delta t$$

Taking the limit as the number of little time intervals grows and their lengths shrink, we obtain the formula for the arc length of the curve:

$$S = \int_a^b \sqrt{(x'(t))^2 + (y'(t))^2 + (z'(t))^2}\,dt = \int_a^b |\mathbf{v}(t)|\,dt$$

The values of a and b determine where we begin and end on the curve. We start measuring length at $\mathbf{r}(a)$ and stop at $\mathbf{r}(b)$.

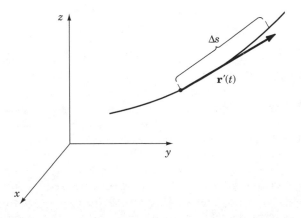

Figure 6.4 Δs is approximately equal to $|\mathbf{r}'(t)|\,\Delta t$.

Example 3   Our bumblebee drinks up some fermented nectar at coordinates $(0, 1, 0)$ and heads back to its hive. At time t its position is given by

$$\mathbf{r}(t) = \langle \sin t, \; \cos t, \; t \rangle$$

with units in feet. The trip back to the hive takes 4π seconds. How far does the bee travel?

    Solution   Plugging $t = 0$ into $\mathbf{r}(t)$, we see that $\mathbf{r}(0) = \langle 0, 1, 0 \rangle$. So time $t = 0$ corresponds to when the bee is at its favorite flower. We calculate that

$$|\mathbf{v}(t)| = \sqrt{(x'(t))^2 + (y'(t))^2 + (z'(t))^2} = \sqrt{(\cos t)^2 + (-\sin t)^2 + (1)^2} = \sqrt{2}$$

So the arc length that the bee travels is

$$S = \int_a^b |\mathbf{v}(t)| \, dt = \int_0^{4\pi} \sqrt{2} \, dt = \sqrt{2} t \Big|_0^{4\pi} = (\sqrt{2})(4\pi)$$

The flight path length is $4\sqrt{2}\pi$, or about 17.77 feet.

    Hey, whatever sweetens your honey.

6.2   Curvature

Curvature is a measurement of how much a curve is bending. Little circles may not be big, but they have big curvature. Great large circles going all the way around the world have very small curvature. Microscopic circles too tiny to see with the naked eye have super big curvature. Given a random curve in space, the curvature can vary as we move along the curve. We're going to see how to measure exactly how much curvature there is at a point on a parameterized curve like $\mathbf{r}(t) = \langle x(t), y(t), z(t) \rangle$.

    Curvature is measured by seeing how much the tangent vector changes. Now the tangent vector of $\mathbf{r}(t)$ can change in two ways. It can point in different directions as we move along the curve, which is what happens as we move around a circle. It can also change by getting shorter or longer. This is not the kind of change we care about here, since it doesn't reflect the shape of the curve, but rather how fast we are moving along the curve. So to measure curvature, we choose a parametrization $\mathbf{r}(s)$ in which $|\mathbf{r}'(s)| = 1$. This is called *parametrizing a curve by arc length*.

    If we're all riding in a car and you are driving, the acceleration can affect us in two ways. If you constantly hit the brake and accelerator, then we'll get thrown backward and forward, even if the road is straight as a ruler. This is not measuring curvature of the road; rather it's measuring that you don't have the faintest idea how to drive—perhaps you got your license from a correspondence school. On the other hand, you might have the car on cruise control, with the speedometer locked at 55 mph, and we're driving on a curvy mountain road. The acceleration you feel then, which throws us violently from side to side, is measuring the curvature of the road.

Group Project   Try both of the above driving techniques with a car full of your fellow students, after a big meal. Be sure to provide airsickness bags, and see which method causes the most nausea.

Example 1   Show that

$$\mathbf{r}(s) = \left\langle R\cos\frac{s}{R}, R\sin\frac{s}{R}, 0 \right\rangle$$

is a parametrization by arc length of the circle of radius R in the xy-plane.

Solution   First let's see that this gives the circle of radius R. Since

$$z = 0$$

the curve lies in the xy-plane, and since

$$\left(R\cos\frac{s}{R}\right)^2 + \left(R\sin\frac{s}{R}\right)^2 = R^2\left(\cos^2\frac{s}{R} + \sin^2\frac{s}{R}\right) = R^2$$

all the points on this curve satisfy $x^2 + y^2 = R^2$, the equation for the circle of radius R.

Now we need to check that the velocity vector has length equal to 1. So let's calculate its length:

$$\mathbf{r}'(s) = \left\langle -R\frac{1}{R}\sin\frac{s}{R}, R\frac{1}{R}\cos\frac{s}{R}, 0 \right\rangle = \left\langle -\sin\frac{s}{R}, \cos\frac{s}{R}, 0 \right\rangle$$

and

$$|\mathbf{r}'(s)| = \sqrt{\sin^2\frac{s}{R} + \cos^2\frac{s}{R}} = \sqrt{1} = 1$$

So the velocity vector has length 1 and we have a parametrization by arc length.

With a curve parametrized by arc length, it's only the rate of change of direction that accounts for the rate of change of $\mathbf{r}'(s)$. But the rate of change of direction is the rate of change of the first derivative, which is the second derivative, and the size of the second derivative is what we used to measure curvature.

We call a curve parametrized by arc length a *unit speed curve* since its speed is always constantly equal to 1.

The curvature $\kappa(s)$ of a unit speed curve $\mathbf{r}(s)$ is given by $\kappa(s) = |\mathbf{r}''(s)|$.

Make sure you have a curve parametrized by arc length when you use this formula.

Example 2 Show that the curvature of the circle of radius R is $1/R$.

Solution From the above example, we know that the curve given by $\mathbf{r}(s) = \langle R \cos(s/R), R \sin(s/R), 0 \rangle$ is a unit speed curve and gives a parametrization of the radius R circle by arc length. Then

$$\mathbf{r}'(s) = \left\langle -\sin\left(\frac{s}{R}\right), \cos\left(\frac{s}{R}\right), 0 \right\rangle$$

and

$$\mathbf{r}''(s) = \left\langle -\frac{1}{R}\cos\frac{s}{R}, -\frac{1}{R}\sin\frac{s}{R}, 0 \right\rangle$$

$$\kappa(s) = |\mathbf{r}''(s)| = \sqrt{\left(-\frac{1}{R}\cos\frac{s}{R}\right)^2 + \left(-\frac{1}{R}\sin\frac{s}{R}\right)^2} = \sqrt{\left(\frac{1}{R}\right)^2} = \frac{1}{R}$$

A circle of radius R has curvature $1/R$, which matches what we said about big circles having small curvature and small circles having big curvature.

But what if we don't have a parametrization by arc length? These parametrizations can be awfully difficult to find for our favorite curve. Fortunately, there's a simple formula we can use to compute curvature which adjusts for curves given by parametrizations that are not unit speed.

Since the vector-valued function $\mathbf{r}'(t)/|\mathbf{r}'(t)|$ is a unit vector that is tangent to the curve at all points, it is often called the *unit tangent vector* and it is denoted $\mathbf{T}$. Then we have the formula for computing the curvature of any parametrized curve:

The curvature of an arbitrary curve $\mathbf{r}(t)$ is given by $\kappa(t) = \dfrac{|\mathbf{T}'(t)|}{|\mathbf{r}'(t)|}$.

Example 3   Find the curvature of $y = x^2$ at the point $(x, y) = (1, 1)$.

Solution   Don't worry yet about unit speed parametrizations. What about any parametrization at all? Since $y = x^2$ on the curve, we can take the parametrization

$$x = t$$
$$y = t^2$$

Then it is certainly true that for any value of t, $y = x^2$, so we take $\mathbf{r}(t) = \langle t, t^2 \rangle$ as our parametrized curve, lying in the xy-plane.

Since $\mathbf{r}'(t) = \langle 1, 2t \rangle$,

$$|\mathbf{r}'(t)| = \sqrt{(1)^2 + (2t)^2} = \sqrt{1 + 4t^2}$$

So

$$\mathbf{T}(t) = \frac{\mathbf{r}'(t)}{|\mathbf{r}'(t)|} = \frac{\langle 1, 2t \rangle}{\sqrt{1 + 4t^2}} = \left\langle \frac{1}{\sqrt{1 + 4t^2}}, \frac{2t}{\sqrt{1 + 4t^2}} \right\rangle$$

Then, taking derivatives, we have

$$\mathbf{T}'(t) = \left\langle \frac{-4t}{(1 + 4t^2)^{3/2}}, \frac{2}{(1 + 4t^2)^{3/2}} \right\rangle$$

$$|\mathbf{T}'(t)| = \sqrt{\left[\frac{-4t}{(1 + 4t^2)^{3/2}} \right]^2 + \left[\frac{2}{(1 + 4t^2)^{3/2}} \right]^2}$$

$$= \sqrt{\frac{16t^2 + 4}{(1 + 4t^2)^3}} = \sqrt{\frac{4}{(1 + 4t^2)^2}} = \frac{2}{1 + 4t^2}$$

So

$$\kappa(t) = \frac{|\mathbf{T}'(t)|}{|\mathbf{r}'(t)|} = \frac{2/(1 + 4t^2)}{\sqrt{1 + 4t^2}} = \frac{2}{(1 + 4t^2)^{3/2}}$$

When $t = 1$, and we are at the point $(1, 1)$ on the curve, we have curvature

$$\kappa(1) = \frac{2}{(5)^{3/2}} \approx 0.1789$$

Just in case you were wondering.

6.3 Velocity and acceleration

Let's face it. "Acceleration" is just a cool word. People try to use it to impress other people. As in, "I have accelerated my consumption of carbohydrates," which is a fancy way of saying, "I feel like a pig." Mathematicians and physicists like the word, too. So they use it to help describe the motion of a particle.

As we've seen, the tangent vector $\mathbf{r}'(t)$ of a moving particle $\mathbf{r}(t)$ gives its velocity, and the magnitude $|\mathbf{r}'(t)|$ gives its speed. The *acceleration* of the particle is the derivative of its velocity vector, and it too is a vector:

$$\mathbf{a}(t) = \mathbf{v}'(t) = \mathbf{r}''(t)$$

The acceleration is the second derivative of the position vector $\mathbf{r}(t)$.

Example 1 Helga, the designated driver, is taking home her three roommates, who have each had one too many, and who are sprawled all over the back-seat. If the car's acceleration ever has a magnitude of more than 750 mph^2, then there will be a pool of recycled beer in the car. The path taken by Helga's car is given by the parametrized curve

$$\mathbf{r}(t) = \langle 30 \sin t, 30 \cos t, 10t \rangle$$

Will Helga have an urgent need for car freshener the next morning?

Solution Helga's car's velocity is given by $\mathbf{r}'(t) = \langle 30 \cos t, -30 \sin t, 10 \rangle$ and the acceleration is given by $\mathbf{r}''(t) = \langle -900 \sin t, -900 \cos t, 0 \rangle$. The magnitude of the acceleration is

$$|\mathbf{r}''(t)| = \sqrt{(-900 \sin t)^2 + (-900 \cos t)^2 + 0^2} = 900 \sqrt{\sin^2 t + \cos^2 t} = 900$$

Looks like Helga better get out her rubber gloves and sponges. (Actually, it's not that surprising her roommates didn't feel well. This is the parametrized curve of a helix. Turns out she got confused and was driving in a parking garage.)

Example 2 Suppose you hurl a wedge of Gouda cheese from the corner of the roof of a building 50 feet above the ground. You launch the cheese with an initial velocity vector $\mathbf{v}_0 = \langle 8, 6, 4 \rangle$ in feet per second. Your roommate is standing on the ground at a point 14 feet 4 inches in the x-direction and 10 feet 9 inches

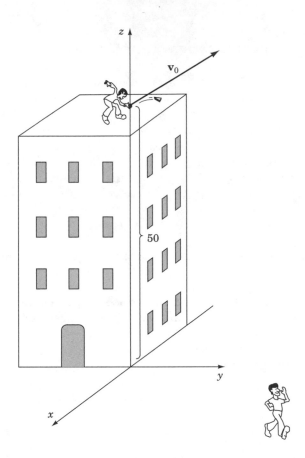

Figure 6.5 Will the cheese and the roommate connect?

in the *y*-direction from the corner of the building. Determine if your cheese will hit him. (See Figure 6.5.)

Solution This is called a *projectile problem,* where the cheese is playing the role of the projectile and your roommate is the projectile attractor. Even if the cheese were replaced by a bag of raisins or a partially melted pint of Cherry Garcia ice cream, we would still call this a projectile problem.

In a situation like this all of the acceleration is coming from gravity. The acceleration due to gravity near the surface of the earth is always 32 feet per second squared. That's in the downward vertical direction. So the acceleration vector is given by $\mathbf{a} = \langle 0, 0, -32 \rangle$.

Integrating this vector gives the velocity vector

$$\mathbf{v}(t) = \langle 0, 0, -32t \rangle + \mathbf{C}$$

Notice that we pick up an arbitrary vector $\mathbf{C}$. But this velocity vector $\mathbf{v}(t)$ must match with the inital velocity vector when $t = 0$. So

$$\langle 8, 6, 4 \rangle = \mathbf{v}_0 = \mathbf{v}(0) = \langle 0, 0, 0 \rangle + \mathbf{C}$$

So $\mathbf{C} = \langle 8, 6, 4 \rangle$. Plugging this into $\mathbf{v}(t)$, we find

$$\mathbf{v}(t) = \langle 8, 6, 4 - 32t \rangle$$

Now, we find the position vector by integrating this velocity vector:

$$\mathbf{r}(t) = \langle 8t, 6t, 4t - 16t^2 \rangle + \mathbf{D}$$

Again, we have an arbitrary vector $\mathbf{D}$ that gets tacked on when we integrate. But we do know that the initial position of the cheese is at coordinate $\mathbf{r}_0 = \langle 0, 0, 50 \rangle$.

So the position vector must match with this when $t = 0$. That is,

$$\langle 0, 0, 50 \rangle = \mathbf{r}_0 = \mathbf{r}(0) = \langle 0, 0, 0 \rangle + \mathbf{D}$$

So $\mathbf{D} = \langle 0, 0, 50 \rangle$. Plugging that into $\mathbf{r}(t)$ yields

$$\mathbf{r}(t) = \langle 8t, 6t, 50 + 4t - 16t^2 \rangle$$

We now know the position of the cheese from when it is launched to when it reaches a height of 6 feet above the ground, which happens to be the height of your roommate. So we would like to know what its x- and y-coordinates are when it has z-coordinate equal to 6, to decide whether or not it hits your roommate. (Your roommate would also like to know.) So we set the z-coordinate of our position function equal to 6 and solve for t:

$$50 + 4t - 16t^2 = 6$$
$$4t^2 - t - 11 = 0$$

By the quadratic formula, this occurs when $t \approx 1.788$ seconds.

Plugging this value for t into $\mathbf{r}$, we find that $\mathbf{r}(1.788) \approx \langle 14.30, 10.73, 6 \rangle$. Given where our roommate is positioned, we can rest assured that the cheese will strike him, at least on the upper torso if not the head.

Common Mistake   In this last example, the vectors $\mathbf{C}$ and $\mathbf{D}$ turned out to just equal the initial vectors $\mathbf{v}_0$ and $\mathbf{r}_0$. However, that only happens when the rest of $\mathbf{v}(t)$ and $\mathbf{r}(t)$ are 0 at time $t = 0$. It is wrong to assume that this always happens. Example 3 illustrates the problem.

Example 3   Given a velocity vector $\mathbf{v}(t) = \langle \sin t, \; \cos t, \; t \rangle$ and initial position vector $\mathbf{r}_0 = \langle 1, 1, 1 \rangle$, find the general position vector $\mathbf{r}$.

Solution   Integrating $\mathbf{v}$, we obtain

$$\mathbf{r}(t) = \left\langle -\cos t, \; \sin t, \; \frac{t^2}{2} \right\rangle + \mathbf{C}$$

Since the intial postion vector must equal the general position vector at time $t = 0$, we have

$$\langle 1, 1, 1 \rangle = \mathbf{r}_0 = \mathbf{r}(0) = \langle -1, 0, 0 \rangle + \mathbf{C}$$

So $\mathbf{C} = \langle 2, 1, 1 \rangle$, rather than being equal to $\mathbf{r}_0$.
Plugging this in for $\mathbf{C}$, we get

$$\mathbf{r}(t) = \left\langle 2 - \cos t, \; 1 + \sin t, \; 1 + \frac{t^2}{2} \right\rangle$$

which is the position vector for any time t.

Hey, whatever lubes your chassis.

Surfaces and Graphing

Let's begin with a quick review of some simple types of curves that occur in the plane.

7.1 Curves in the plane: a retrospective

1. First there is the *parabola*, the generic form of which is $y = x^2$ (see Figure 7.1a).

We can adjust it up or down, as in $y = x^2 + 1$ and $y = x^2 - 2$. We can widen or thin it down, as in $y = x^2/2$ and $y = 3x^2$ (Figure 7.1b).

We can flip it as in $y = -x^2$ (Figure 7.1c), or move it so it opens on the x-axis as in $x = y^2$ and $x = -y^2$ (Figure 7.1d).

2. Then there is the *ellipse*, which is given by an equation of the form $x^2/a^2 + y^2/b^2 = 1$. This is a squashed circle hitting the x-axis at a and $-a$ and the y-axis at b and $-b$, as in Figure 7.2. (When $a = b$, it is a circle.)

3. And then there is the *hyperbola*. This has an equation like $y^2 - x^2 = 4$. Unlike the ellipse or the circle, the signs in front of the x^2 and the y^2 terms are

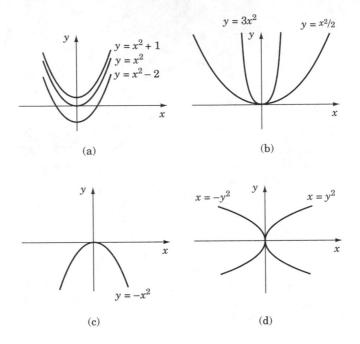

Figure 7.1 A variety of parabolas.

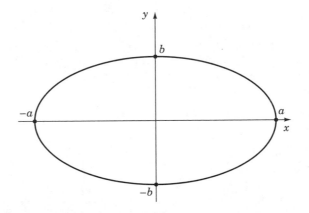

Figure 7.2 The ellipse given by $x^2/a^2 + y^2/b^2 = 1$.

opposite. In this example, note that when $x = 0, y = 2$ or -2. So the resulting hyperbola will intersect the y-axis twice, as in Figure 7.3a.

Similarly, $x^2 - y^2 = 9$ will intersect the x-axis at 3 and -3, as in Figure 7.3b. Constants in front of the x^2 and y^2 terms will change how quickly the hyperbola curves, but qualitatively leave it unchanged.

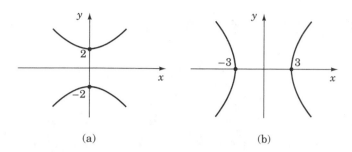

(a)   (b)

Figure 7.3 Hyperbolas.

7.2 Graphs of equations in 3-space

Psychologists use an amazing array of tests to decide whose nuts are loose. One of the most famous is the Rorschach test. This is where they show you an inkblot and ask you what it resembles. If you say, "That is a lovely butterfly flitting from flower to flower on a sunny spring day," then you're fine. If you say, "That is my father dancing the tango with our neighbor's poodle," they strap you to a gurney and pump you full of thorazine.

In this section, we will derive our own Rorschach test. But our test will be based on surfaces in 3-space. This has the definite advantage that a surface is unlikely to look like your father dancing with the neighbor's poodle. On the other hand, it does mean we will need to become proficient at graphing and recognizing various surfaces in 3-space. Now what exactly do we mean by graphing a surface in 3-space? Let's look at what is probably the simplest example.

Example 1 Graph all the points (x, y, z) in 3-space that satisfy

$$x^2 + y^2 + z^2 - 1 = 0$$

Solution Moving the 1 to the other side of the equation and taking square roots, we have $\sqrt{x^2 + y^2 + z^2} = 1$. This is exactly the set of points that have distance 1 from the origin, so we get the sphere shown in Figure 7.4.

The sphere is not much use for our Rorschach test, since everyone agrees that it represents an alien orb that abducts people for invasive experiments. But a sphere is a good example of a surface. More generally, we will say that the graph of an equation $F(x, y, z) = 0$ is the set of all points (x, y, z) in 3-space that satisfy the equation. Such a graph will be called a *surface*.

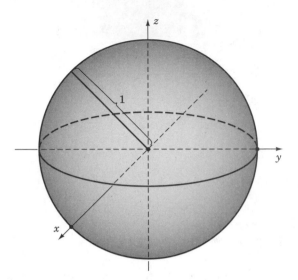

Figure 7.4 The sphere given by $x^2 + y^2 + z^2 - 1 = 0$.

Example 2 Draw the graph of $2x + 3y + 4z = 12$.

Solution Notice that the equation is linear. That is to say it has no square roots, no powers like x^2, no products like xy, and heaven forbid, no sines, cosines, or logs. It is simply $Ax + By + Cz = D$, where A, B, C, and D are constants. An equation of this form is called *linear*, and its graph will always be a plane. Of course, that doesn't answer the question of which plane it is. To figure that out, we will use the patented technique of seeing where the plane intersects the coordinate axes.

Step 1 Where does the plane intersect the x-axis? Well, the x-axis is all the points where $y = 0$ and $z = 0$, so if we set $y = 0$ and $z = 0$ in the equation we will see where the plane intersects the x-axis. We get $2x = 12$, so $x = 6$.

Step 2 Where does it intersect the y-axis? Set $x = 0$ and $z = 0$. We get $3y = 12$, so $y = 4$.

Step 3 Where does it intersect the z-axis? (If you don't see the pattern by now, call your optometrist.) Set $x = y = 0$. Then $4z = 12$, and $z = 3$.

Step 4 Draw the picture. We now know where the plane intersects the x-, y-, and z-axes. We will just shade that part of the plane that occurs in the positive (sometimes called first) octant of space, by connecting these three points by lines to obtain Figure 7.5.

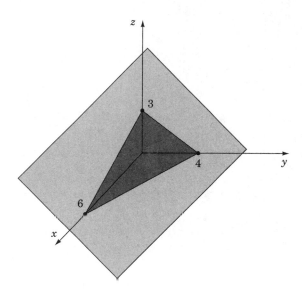

Figure 7.5 The plane given by $2x + 3y + 4z = 12$.

This technique for graphing a plane works well if the plane intersects all three axes at different points. If it doesn't, as for instance occurs if it passes through the origin, then the best idea is to draw the normal vector and then the plane perpendicular to it.

Any time we have a linear equation, we obtain a plane for its graph. But what if we don't have a linear equation? Well, that is what the rest of this section is about.

A VARIABLE IS ON VACATION

Suppose we want to graph an equation in 3-space, but one of the variables is missing from the equation, as in $y^2 + z^2 = 1$. Here, x is on vacation, taking a coffee break, or pretending to be ill, so he can stay home and watch *The Price Is Right*. So it's up to just y and z. Let's figure out what this would look like if $x = 0$, which is to say, if we restrict ourselves to the yz-plane. Then we realize that we have just taken a trip backward in time to single-variable calculus, because now we are just graphing an equation in two variables in a plane. And this particular equation gives a circle of radius 1 centered at the origin. This tells us how our equation is intersecting the yz-plane.

But remember, x is on vacation. And when you're on vacation, anything goes. The usual social strictures are out the window. It's a free-for-all. So there are no restrictions on x. It's spring break in Fort Lauderdale for x. For a point (x, y, z) in the graph of this equation, x can be anything, but its y- and z-coordinates must still satisfy $y^2 + z^2 = 1$. So if $(0, y_0, z_0)$ is in the graph,

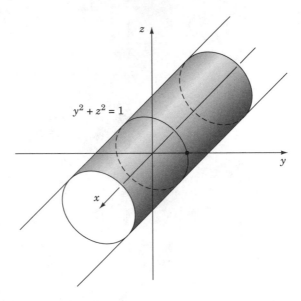

Figure 7.6 The infinite tin can.

so is (x, y_0, z_0) for any x. All the horizontal lines parallel to the x-axis that pass through the circle of radius 1 centered at the origin in the yz-plane are contained in the graph. The resulting graph looks like a traditional cylinder or tin can, sitting on its side and extending in both directions forever, as in Figure 7.6.

Example 3 Graph $z = 1 - x^2$.

Solution This time, y is taking the vacation in Cancun. So in the xz-plane we have an upside-down parabola. Since y can be anything, the surface is obtained by taking all of the lines parallel to the y-axis that pass through this parabola. This is where the Rorschach test comes in. You have to decide what it looks like. Could be an upside-down rain gutter or a train tunnel. Freud would have lots to say about your answer. To us, it is the infinite quonset hut, as in Figure 7.7.

Its official name is the *parabolic cylinder*. Notice the use of the word *cylinder*. In ordinary usage, cylinder means drum-shaped, as in the soup can made famous by Andy Warhol. But in mathematics, a cylinder is any surface obtained by taking a curve in a plane and all of the lines through that curve parallel to a given line not in that plane. So we can have circular cylinders, parabolic cylinders, elliptic cylinders, and hyperbolic cylinders. Really opens up the possibilities, doesn't it?

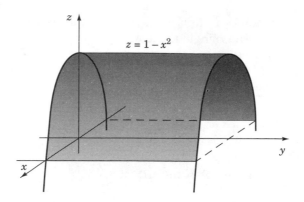

Figure 7.7 The infinite quonset hut.

7.3 Surfaces of revolution

Here's a simple method to generate interesting surfaces. Take a curve in one of the coordinate planes and rotate it about one of the coordinate axes. The result will be a surface in space, called a *surface of revolution*. For instance, suppose we have the curve $z = f(y)$ in the yz-plane and we want to rotate it about the z-axis, as in Figure 7.8.

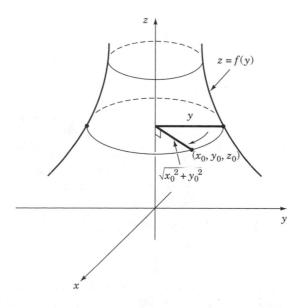

Figure 7.8 Rotating $z = f(y)$ about the z-axis.

Then the distance from a point (y, z) on the curve back to the z-axis is y. When we rotate, that single point will generate an entire circle of points, all that same distance from the z-axis. But for a general point (x_0, y_0, z_0), the distance to the z-axis is $\sqrt{x_0^2 + y_0^2}$. So the new points must satisfy the same equation as the original curve, but with each y occurring in the equation replaced by $\sqrt{x^2 + y^2}$. So

$$z = f(\sqrt{x^2 + y^2})$$

is the equation of the resulting surface.

Example   Find an equation of the surface that results when we revolve the curve $x = -z^2$ about the x-axis.

Solution   This is easy. The distance from a point on the curve to the x-axis is z. So we replace each z in the equation with $\sqrt{y^2 + z^2}$ to obtain

$$x = -(\sqrt{y^2 + z^2})^2$$
$$x = -y^2 - z^2$$

The resulting surface appears in Figure 7.9.

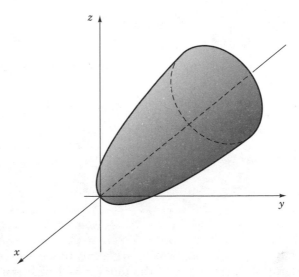

Figure 7.9 Revolving $x = -z^2$ about the x-axis.

7.4 Quadric surfaces (the -oid surfaces)

"-oid" words always refer to things that are weird, uncomfortable, or just plain offensive. Take for instance, paranoid. That's a good one. Or how about steroids. Now that's something to stay away from unless you want to win the blue ribbon in the bull category at the 4-H fair. How about void—not a place where you want to spend a lot of time. And then there's hemorrhoid. No comment.

Now meet a whole new set of "-oid" words. These are not offensive words in the way that the previous ones were. We will have ellipsoid, hyperboloid, paraboloid, and hyperbolic paraboloid. These are the so-called *quadric surfaces*. They're called quadric because all of them are given by equations that are quadric, which is to say, equations that involve first and second powers of the variables but no more. They usually look like

$$ax^2 + by^2 + cz^2 + dx + ey + fz + g = 0$$

Given a specific equation of this form, how do you figure out what the surface looks like? As your grandmother used to always say, "Honey, please eat the toast one slice at a time."

At the time, you thought, "Another one of Grammy's nutty rules." But you eventually found out that this is the way it's done. And now her advice is going to come in handy, because we will understand these surfaces one slice at a time, hold the butter.

Example 1   Graph the surface $z = x^2 + y^2$.

Solution   We slice it with each of the coordinate planes. First take the plane $x = 0$, which is the yz-plane. Setting $x = 0$ in the equation yields $z = y^2$. This is just your generic standard parabola opening along the positive z-axis in the yz-plane.

Now, slice it with the plane $y = 0$. This gives $z = x^2$, another generic parabola, this time in the xz-plane.

Now, slice with $z = 0$, the xy-plane, to obtain $0 = x^2 + y^2$. Since both x^2 and y^2 are nonnegative numbers, they can add up to 0 only if they are both equal to 0. So the only solution to this equation is the single point $(0, 0)$.

We take these three intersections and draw them in the respective co-ordinate planes, obtaining a framework for our surface. It appears to be a bowl-shaped object, suitable for holding nuts or radioactive iodine. But how can we be sure it's a bowl? Let's slice it with a plane which lies above the xy-plane, say $z = 1$. Its intersection with this plane is given by $1 = x^2 + y^2$, a circle of radius 1. Drawing it in the plane $z = 1$ confirms our suspicion that this is a bowl (see Figure 7.10a).

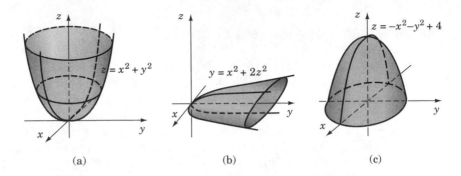

Figure 7.10 A selection of paraboloids.

This type of bowl is a generalization of the parabola, and goes by the name of a *paraboloid*. This particular one is the most generic Wal-Mart version. By putting constants in front of the various terms, and/or changing the roles of x, y, and z, we can create fancier paraboloids (and charge more for them).

For instance, $y = x^2 + 2z^2$ appears in Figure 7.10b. The graph of the equation $z = -x^2 - y^2 + 4$ appears in Figure 7.10c.

Rorschach Test Question 1 Of which of the following does Figure 7.10c remind you?

1. Hill

2. Fish head

3. Tongue

Now for another type of surface.

Example 2 Graph the surface $x^2 + \dfrac{y^2}{9} + \dfrac{z^2}{16} = 1$.

Solution When we slice with the coordinate planes, we obtain an ellipse in each case:

$x = 0$ gives the ellipse $\dfrac{y^2}{9} + \dfrac{z^2}{16} = 1$.

$y = 0$ gives the ellipse $x^2 + \dfrac{z^2}{16} = 1$.

$z = 0$ gives the ellipse $x^2 + \dfrac{y^2}{9} = 1$.

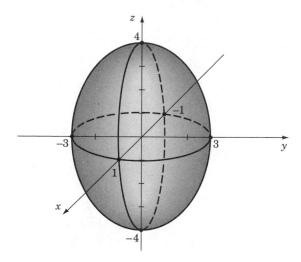

Figure 7.11 The ellipsoid $x^2 + \dfrac{y^2}{9} + \dfrac{z^2}{16} = 1$.

Graphing each of these in their respective coordinate planes, we see the framework for a squashed sphere, called an *ellipsoid*, in Figure 7.11.

Rorschach Test Question 2 Of which of the following does Figure 7.11 remind you?

1. Dinosaur egg

2. Tylenol pill

3. Gonad

Let's look at another type.

Example 3 Graph $z^2 = x^2 + 4y^2 - 4$.

Solution Slicing with $x = 0$ gives the hyperbola $z^2 = 4y^2 - 4$. When $z = 0$, the hyperbola hits the y-axis at ± 1, but when $y = 0$, there are no solutions. So it misses the z-axis.

Slicing with $y = 0$ gives the hyperbola $z^2 = x^2 - 4$. And slicing with $z = 0$ gives the ellipse $0 = x^2 + 4y^2 - 4$. This generates a framework as in Figure 7.12, which we can complete to the surface shown. We can slice this with planes at $z = 1, 2, -1, -2$ to confirm that it looks the way we think it does.

This is called a *one-sheeted hyperboloid*.

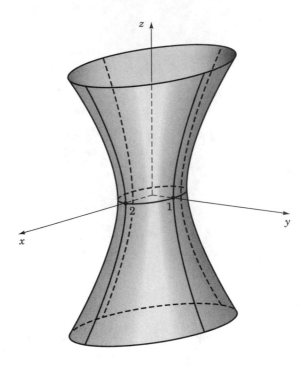

Figure 7.12 The one-sheeted hyperboloid given by $z^2 = x^2 + 4y^2 - 4$.

Rorschach Test Question 3 Of which of the following does Figure 7.12 remind you?

1. Nuclear cooling tower

2. Corset

3. Your uncle in a corset

Example 4 Graph $z^2 = x^2 + y^2$.

Solution. When we slice with $x = 0$, we get $z^2 = y^2$, or $z = \pm y$. This is a pair of crossed lines. Similarly, $y = 0$ produces another pair of crossed lines given by $z = \pm x$. Finally, $z = 0$ gives $x^2 + y^2 = 0$, which has only the solution $(x, y) = (0, 0)$. After drawing the framework for this, it is still very difficult to see the corresponding surface.

So let's slice with $z = 1$. This gives a circle of radius 1. Similarly, $z = 2$ gives a circle of radius $\sqrt{2}$. The planes $z = -1$ and $z = -2$ yield the same pair

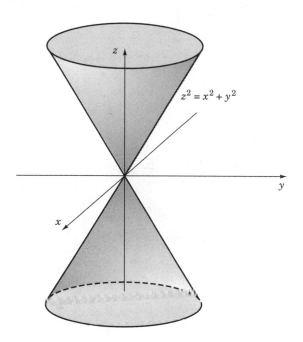

$$z^2 = x^2 + y^2$$

Figure 7.13 When two cones meet.

of circles. Now it is apparent what is going on. We have a pair of cones meeting at the origin, as in Figure 7.13.

Example 5   Graph $z^2 = x^2 + y^2 + 1$.

Solution   By now, you can do the slicing and dicing. This surface is a *hyperboloid of two sheets,* as in Figure 7.14.

Example 6   Graph $z = x^2 - y^2$.

Solution   This last one is worth doing carefully. Slicing with the $x = 0$ plane, we find $z = -y^2$, a downturned parabola. Slicing with the $y = 0$ plane we find $z = x^2$, an upturned parabola. The framework for this surface so far is two parabolas that have backed into one another so they share one point and they meet perpendicularly at that point.

Slicing with the $z = 0$ plane, we have $0 = x^2 - y^2$, or $y = \pm x$. This is two crossed lines. The framework appears in Figure 7.15a. The surface itself appears in Figure 7.15b. It is a saddle or a Pringle's potato chip.

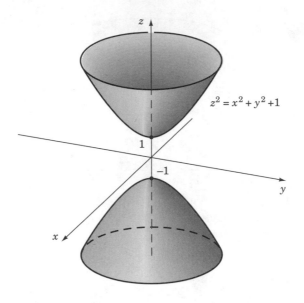

Figure 7.14 A hyperboloid of two sheets.

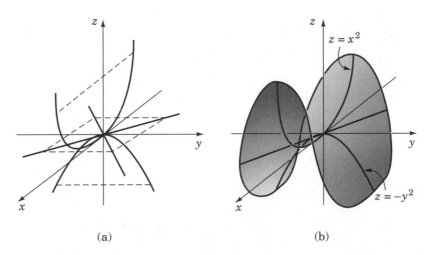

(a) (b)

Figure 7.15 The hyperbolic paraboloid given by $z = x^2 - y^2$.

Mathematicians like to call it the *hyperbolic paraboloid.* Makes it sound better than an old piece of leather tied around a horse.

Rorschach Test Question 4   Of which of the following does Figure 7.15b remind you?

1. Pringle's potato chip

2. Saddle

3. Underside of a cowboy's jeans

Okay, we are ready to see what the Rorschach test says about you. Total the numbers corresponding to your answers to the four questions. Here is a key for determining how you did:

  0–3: You have trouble following instructions.

  4–6: You are absolutely certifiably normal.

  7–9: Have you considered counseling?

10–11: A short stay in a well-padded room might be appropriate.

  12: You are a walking time bomb. Call the nearest high security facility and have them come pick you up.

  13: You are in the wrong math course. Addition is a prerequisite for calculus.

Functions of Several Variables and Their Partial Derivatives

8.1 Functions of several variables

So far in calculus, we have enjoyed meeting functions of a single variable, like $f(x)$. They are simple and straightforward. Nothing fancy to operate. We just put in a single number and wait for the output. We know them, and some of them we even love. Take $f(x) = \sin x$, for example. What's not to love? Even the function $\mathbf{r}(t) = \langle 2 + t, \, 3 - 5t, \, 7t \rangle$ only depends on a single variable t, though it gives us a vector in 3-space as its output.

But we're moving up in the world—we'd like a penthouse apartment with a park view, a red convertible, and we'd like to keep some fancier functions of more than one variable around to make the place look classy.

What's a way to think of a function of two variables? It's just a function machine with two mouths. We feed f two numbers, x and y, and it spits out another number, $f(x, y)$. Why does it spit out just one number? Why not two, or even three? Well it could. But first, let's get a handle on this crowd. The hardcore stuff comes later.

Example 1 $f(x, y) = 2x + y$

Put in the pair of numbers (2, 3). What pops out? This is not a trick question; 7 pops out. Does it matter which order the pair was put in? Absolutely. Put in the pair (3, 2). If you still get 7, have a cup of coffee and try again. You should get 8.

We can think of the number $f(x, y)$ as telling us a height z above the point (x, y) in the xy-plane. In other words, we graph the function by setting $z = f(x, y)$ and graphing all the resulting points (x, y, z). The whole collection of points (x, y, z) describes a surface in three-dimensional space. In our example, we graph all of the points (x, y, z) that satisfy $z = 2x + y$. We call z the *dependent variable*, and x and y the *independent variables*. So x and y can claim z as a dependent on their tax returns. We get the surface that appears in Figure 8.1.

This is called a *linear function* because its graph, the surface in Figure 8.1, is like a line. Uh, well, okay, it's not a line, but it's as close as you can get if you're a surface; it's a plane. We can have linear functions in any number of variables. Here's one in four variables:

$$f(x, y, z, w) = 2x - y + 57z - \sqrt{2}w + 3$$

Picturing this one successfully would require a tour of five-dimensional space, which we're not up for right now. To qualify f as a linear function,

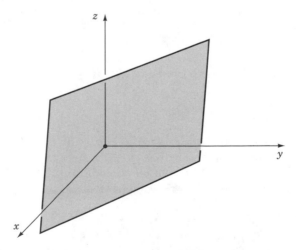

Figure 8.1 The graph of $f(x, y) = 2x + y$, a linear function.

there can't be any funny business in the formula for f. No squares, no cosines, no square roots, no stray cats. Each term has to be just your classic (constant) $\times$ (variable). Although there is a $\sqrt{2}$ in the formula, that's okay as $\sqrt{2}$ is a constant. But we could not have a $\sqrt{w}$. Linear functions are some of the simplest functions that exist. They are the vanilla ice cream of the function business.

For those that prefer their ice cream with chocolate sauce and a cherry, we have Example 2.

Example 2 $f(x, y) = x^2 + y^2 - 2y$

Definitely not a linear function, no matter how you slice it. But it should remind you of something, if ever so faintly. That nose, those eyes ... It's got that paraboloid look to it. Except for that ear coming out of the forehead, the $-2y$ term. Let's complete the square on the y terms and see if that helps.

$$z = x^2 + (y^2 - 2y + 1) - 1$$
$$= x^2 + (y - 1)^2 - 1$$

This looks just like the equation of our generic Wal-Mart paraboloid, but the y has been replaced by a $y - 1$ and there is an extra -1 at the end. The fact there is a $y - 1$ instead of a y shifts the paraboloid over one unit in the positive y-direction. The extra -1 shifts the paraboloid down one unit in the z-direction. The resulting surface appears in Figure 8.2.

For those who prefer their ice cream with anchovies and horseradish, there's Example 3.

Example 3 $f(x, y) = \sin x \cos y$

Let's graph all the points (x, y, z) that satisfy $z = \sin x \cos y$. Graphing this is not as hard as it looks. The product $\sin x \cos y$ is zero in a lot of spots. Whenever x is a multiple of π, or y is an odd multiple of $\pi/2$, $f(x, y)$ is zero. A picture in the xy-plane of all the places the graph of this function never gets off the ground appears in Figure 8.3.

What is it doing the rest of the time? Well, imagine trying to make the surface, the graph of the function, out of a sheet of rubber. What we've just learned is that step 1 in this process is to nail the sheet down along all of those grid lines. Doesn't leave you much room for creative canoodling, does

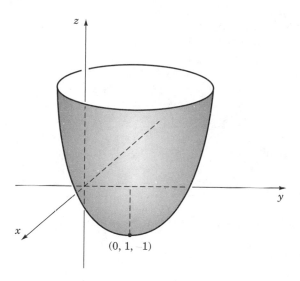

Figure 8.2 The graph of $f(x, y) = x^2 + y^2 - 2y$.

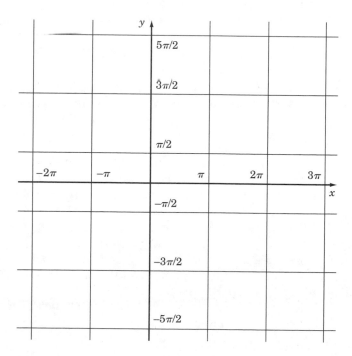

Figure 8.3 The (x, y) values where $\sin x \cos y = 0$ form a grid of lines in the plane.

Figure 8.4 The graph of $f(x, y) = \sin x \cos y$.

it? What happens to the rubber sheet in between the grid lines? Basically, it either goes up, and forms a little hill, or it goes down and forms a charming valley. (See Figure 8.4.)

We can tell which goes where by checking the value of the function at some clever point, like the middle of one of the squares:

$$f(\pi/2, 0) = \sin(\pi/2) \cos(0) = 1$$

So in the square that contains the point $(\pi/2, 0)$ there's a little hill. The hills and valleys alternate like the white and black squares of a chessboard.

Functions in more than one variable are around because, let's face it, life is just more complicated than can be handled by a single variable. If whether you go home for Thanksgiving depends only on how much studying you need to do, the whole question would be really straightforward. Too much work, no turkey. Take-home exams, turkey. But a whole variety of other factors come into play here—whether your obnoxious cousin will be there, how much money you owe your big brother, whether or not you can face another year of sweet potatoes with marshmallows on top. Your decision about hitting the old homestead is a *function of several variables.*

The rest of this chapter is about functions of more than one variable, how to visualize them, how to take limits of them, and yes, since this is calculus, how to differentiate them, and why you would want to do that anyway.

Hey, whatever stuffs your turkey.

Contour curves

In this section, we come to a concept familiar to you outdoorsy types who like to sleep on a bed of rock while mosquitoes drone in your ears. That is the concept of contour curves. When they appear on a contour map, they allow you to visualize the terrain. For our purposes, they allow us to visualize graphs of surfaces in 3-space.

A WALK IN THE WOODS

You've been lost in Tepid Water State Park in the Less-than-Grand Tetons for the last 3 days. You've eaten the leather from your wallet, and sucked the juice out of the last piece of bark your stomach will tolerate, when you remember you have a map in your left front shirt pocket. It's one of those contour maps, with lots of squiggly green lines, each of which represents an altitude. You spot the highest point on the map, Mount Middling, and then peer through the forest to spot the highest point in sight from where you are. Aha, this must be Mount Middling. Before noon the next day, you are perched atop the mountain, looking down at the less-than-awe-inspiring scene. Now the question is how do you get down to the trading post at the park entrance, where they sell prewrapped hotdogs that can be nuked lukewarm in their microwave oven. Of course, you realize that the contour curves represent the intersection curves of horizontal planes with the mountain. So you can reconstruct the shape of the mountain from the map, and find your way to postmodern civilization, hold the mustard.

We do the same thing without the hotdogs. By keeping track of the contour curves of a surface, we can get a good idea of what the surface looks like.

Example 1   Determine contour curves for the surface given by $f(x, y) = 4 - x^2 - y^2$.

Solution   The function $f(x, y) = 4 - x^2 - y^2$ has a graph equal to all the points (x, y, z) in 3-space that satisfy

$$z = 4 - x^2 - y^2$$

We will slice the surface with horizontal planes corresponding to particular constant values for z. When $z = 0$, we get the circle $x^2 + y^2 = 4$ which has radius 2. When $z = 1$, we get the circle $x^2 + y^2 = 3$ with radius $\sqrt{3}$. When $z = 2$, we get a circle of radius $\sqrt{2}$. When $z = 3$, we get a circle of radius 1. And when $z = 4$, we get $x^2 + y^2 = 0$, which is a circle of radius 0, better

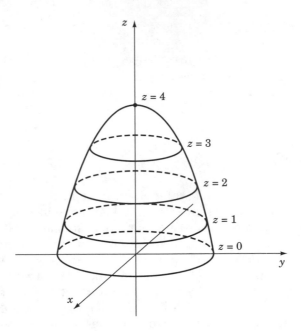

Figure 8.5 The graph of $f(x, y) = 4 - x^2 - y^2$.

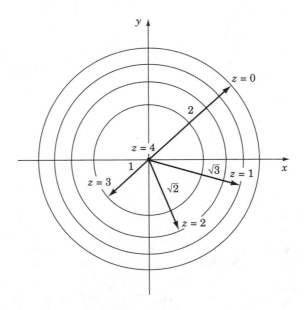

Figure 8.6 The level curves of $f(x, y) = 4 - x^2 - y^2$.

known as a point. Putting these together, as in Figure 8.5, we get our old friend the paraboloid, though it's upside down.

If we project all these intersection curves to the xy-plane and draw them there, labeled by their z-coordinates, we get Figure 8.6. The result is often called a set of *level curves* for the function.

We can work the other way as well.

Example 2   The level curves in Figure 8.7 represent a surface. What kind of surface is it?

Solution   From the level sets, we see that the surface is intersecting horizontal planes above the xy-plane in hyperbolas missing the yz-plane. It is intersecting horizontal planes below the xy-plane in hyperbolas missing the xz-plane. And it intersects the xy-plane in two crossing diagonal lines. There's only one surface that does that, the surface that's close to every cowpoke's heart (actually, the heart is not the body part that it is closest to), and that's a saddle. The simplest such saddle is given by $z = x^2 - y^2$, as in Figure 8.8.

Hey, whatever makes your parakeet twitter.

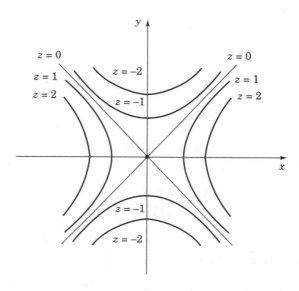

Figure 8.7 The level curves of some surface.

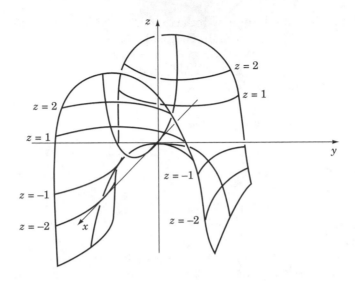

Figure 8.8 A saddle given by $z = x^2 - y^2$.

8.3 Limits

In order to define derivatives of functions of more than one variable, we need to understand their limits. But maybe first we should review limits of functions of one variable.

LIMITS OF FUNCTIONS OF ONE VARIABLE: A RETROSPECTIVE

Do you remember the idea of limits? Basically, limits are about falling in love. We say $\lim_{x \to a} f(x) = L$ if, as x approaches a, $f(x)$ approaches L. That is to say, $f(x)$ becomes more and more enamored of L as x approaches a. It doesn't matter what $f(x)$ equals when $x = a$, it just matters what it's approaching.

For instance,

$$\lim_{x \to 2} x^2 - 3x + 4 = 2^2 - 3(2) + 4 = 2$$

As another example, look at

$$\lim_{x \to 1} \frac{x^2 - 1}{x - 1}$$

Although $\dfrac{x^2 - 1}{x - 1}$ is undefined at $x = 1$, it's still true that

$$\lim_{x \to 1} \frac{x^2 - 1}{x - 1} = \lim_{x \to 1} \frac{(x - 1)(x + 1)}{x - 1} = \lim_{x \to 1} x + 1 = 2$$

With a function of one variable, there are exactly two ways that x can approach a, from the left (so $x < a$) or from the right (so $x > a$). Only in the case that $f(x)$ approaches the same value L when x approaches a from the left and right does the limit exist.

LIMITS OF FUNCTIONS OF MORE THAN ONE VARIABLE

For functions of more than one variable, limits are still about falling in love. Now when you really fall in love, it's not instantaneous. Those sudden infatuations when you see someone across the room usually lead to nothing but trouble, because, no attractive as someone might seem at a quick glance, if he trims his toenails with a machete, and burps himself to sleep at night, your life together will not be idyllic.

No, falling in love should be a slow, gradual process, where each little thing this person does endears him or her to you just a bit more. That's the way functions fall in love.

Example 1   Find the limit of $f(x, y) = x^2y + x$ as (x, y) approaches $(1,1)$.

Solution   As (x, y) approaches $(1,1)$, $x^2y + x$ approaches 2. It doesn't matter what path (x, y) takes to approach $(1,1)$. When it's time to make the commitment, $x^2y + x$ does the deed, follows her heart and marries the man she was meant for, 2. So we write

$$\lim_{(x,y) \to (1,1)} x^2y + x = 2$$

In fact, all the polynomial functions are the marrying kind. If $P(x, y)$ is a polynomial in x and y, then

$$\lim_{(x,y) \to (a,b)} P(x, y) = P(a, b)$$

That is to say, we just evaluate $P(x, y)$ at (a, b) to find the limit. We just plug in. This is because limits of functions of more than one variable obey the same rules that limits of functions of one variable obey. When f and g each have nicely behaving limits, the rules tell us that:

1. $\lim (f + g) = \lim f + \lim g$

2. $\lim cf = c \lim f$

3. $\lim (fg) = \lim f \lim g$

4. $\lim \left(\dfrac{f}{g} \right) = \dfrac{\lim f}{\lim g}$ (if $\lim g \neq 0$)

So

$$
\begin{aligned}
\lim_{(x,y)\to(1,1)} x^2 y + x &= \lim_{(x,y)\to(1,1)} x^2 y + \lim_{(x,y)\to(1,1)} x \\
&= \left(\lim_{(x,y)\to(1,1)} x^2 \right)\left(\lim_{(x,y)\to(1,1)} y \right) + \lim_{(x,y)\to(1,1)} x \\
&= (1)^2(1) + 1 \\
&= 2
\end{aligned}
$$

Limits of functions more general than polynomials are going to be more complicated. Unlike what happens for limits of functions of one variable, there are not just two paths to consider as x approaches a, one from the right and one from the left. No, now we have a point (x, y) in the xy-plane approaching another point (a, b). There are many many different paths from (x, y) to (a, b), as in Figure 8.9, which shows three of them. We could take any of them. We could take a straight line path. We could take a curving path. We could take a spiral path. If the limit is to exist, the answer has to be the same for all of them. You see, true love transcends the path you take to get there. (This book is not just about math. It's about *life*.)

But then the question comes up, how do we check the infinite number of possible paths with which we could reach (a, b)? Let's take a look at an example.

Example 2 Find $\lim\limits_{(x,\,y)\to(0,\,0)} \dfrac{xy}{\sqrt{x^2 + y^2}}$.

Solution Notice that this is an indeterminant form of type $0/0$. Unfortunately, there is no L'Hôpital's rule for functions of more than one variable. So, how do we check every possible path from (x, y) to (a, b)? How about this idea? We will convert to polar coordinates. Converting to polar coordinates is

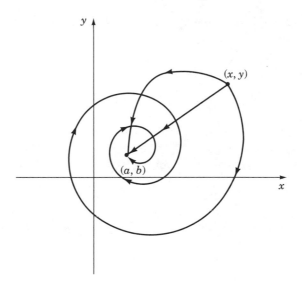

Figure 8.9 Possible paths from (x, y) to (a, b).

a lot like converting to a new religion, only your parents won't be so upset. Remember that

$$x = r \cos \theta \qquad y = r \sin \theta \qquad r = \sqrt{x^2 + y^2}$$

Notice that as (x, y) approaches $(0, 0)$, r approaches 0.

So our limit becomes

$$\lim_{(x,y) \to (0,0)} \frac{xy}{\sqrt{x^2 + y^2}} = \lim_{r \to 0} \frac{(r \cos \theta)(r \sin \theta)}{r} = \lim_{r \to 0} r \cos \theta \sin \theta = 0$$

Since the quantity $\cos \theta \sin \theta$ is never bigger than 1, multiplying it by an r that is shrinking to 0 will shrink the whole expression to 0.

Converting to polar coordinates can work when (x, y) is approaching $(0, 0)$, but won't help otherwise.

Example 3 Determine whether or not this limit exists:

$$\lim_{(x,y) \to (0,0)} \frac{x^2 - y^2}{x^2 + y^2}$$

Solution   If we convert to polar coordinates, we find

$$\lim_{(x,y)\to(0,0)} \frac{x^2 - y^2}{x^2 + y^2} = \lim_{r\to 0} \frac{(r\cos\theta)^2 - (r\sin\theta)^2}{r^2} = \lim_{r\to 0}(\cos\theta)^2 - (\sin\theta)^2$$

This has different values depending on the value of θ. For instance, if $\theta = 0$ [so we are approaching (0,0) along the positive x-axis], this gives 1. If $\theta = \pi/2$ [so we are approaching (0,0) along the positive y-axis], this gives -1. If $\theta = \pi/4$ [so we are approaching (0,0) along the line $x = y$], this gives 0. Since different paths approaching (0,0) result in different values, this limit does not exist!

We could also do this problem without resorting to polar coodinates. Setting $y = 0$ (so we are approaching on the x-axis), the limit becomes

$$\lim_{x\to 0} \frac{x^2 - 0^2}{x^2 + 0^2} = \lim_{x\to 0} 1 = 1$$

But setting $x = 0$ (so we are approaching on the y-axis), the limit becomes:

$$\lim_{x\to 0} \frac{0^2 - y^2}{0^2 + y^2} = \lim_{x\to 0}(-1) = -1$$

Two different paths have given two different results, so the limit DOES NOT EXIST.

Hey, whatever garnishes your potato salad.

8.4 Continuity

Functions come in two varieties, continuous and discontinuous. Continuous functions have nice graphs without holes and tears. Discontinuous functions are dangerous. That's how you recognize them. They generate the kind of terrain that should only be tackled with climbing gear. Suppose you're out for a hike. What should you watch out for? Holes, the bane of every hiker's existence. You put a foot in a hole, and it's bye bye leg bone, and then, you're bear meat come sunset. Of course, it won't do you any good avoiding holes if in the process you walk off the edge of a cliff. Then it's a long, seemingly slow drop to certain death on the rocks below. And the worst part is, your entire life flashes before your eyes, in slo mo, including the time you had the little accident on that bus trip to the Museum of Natural History, and the time Tommie Flecker told everyone that you had a crush on Mr. Rogers. And if holes and cliffs aren't bad enough, the worst is when you walk up a slope and

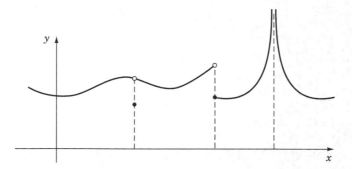

Figure 8.10 A graph with three types of discontinuities.

inadvertently fall into the spewing crater of Mount St. Helens. That can be quite a ride, as the cinder that was once you is spit out into the Oregon sky. Yes, those are all examples of discontinuities.

You probably have fond memories of discontinuities of functions of one variable.

The three situations described, holes, cliffs, and volcanos, each appear in the graph of the function $y = f(x)$ in Figure 8.10, and at each of these points, the function is discontinuous. At the other points, the function is continuous and happy as a clam. We've seen an official definition of what it meant for a function of one variable to be continuous at a point, which goes like this:

A function $f(x)$ is **continuous** at $x = a$ if $\lim_{x \to a} f(x) = f(a)$.

This means three things:

1. The limit exists.

2. The function $f(x)$ is defined at $x = a$.

3. The limit equals $f(a)$.

In the three examples in the graph in Figure 8.10, we see the hole's discontinuity comes from the fact that $\lim_{x \to a} f(x) \neq f(a)$; the cliff from the fact that the left and right limits are different and so the limit doesn't exist; and the volcano from the fact that neither the limit nor $f(a)$ exists.

But now we are talking about functions of two variables. So suppose that $z = f(x, y)$ is a function depending on x and y. Our definition of continuity will be identical.

A function $f(x, y)$ is **continuous** at $(x, y)=(a, b)$ if it is true that $\displaystyle\lim_{(x,y)\to(a,b)} f(x, y) = f(a, b)$.

Again, this really means three things:

1. $\displaystyle\lim_{(x,y)\to(a,b)} f(x,y)$ exists.

2. $f(x, y)$ is defined at (a, b).

3. $\displaystyle\lim_{(x,y)\to(a,b)} f(x,y) = f(a,b)$.

At a point where there is a hole, cliff, or volcano, it will still be the case that $f(x, y)$ is discontinuous, as in Figure 8.11.

We say a function is *continuous everywhere* if it is continuous at all points (a, b) in the xy-plane. Many functions are continuous everywhere. For instance, our friends the polynomials are continuous everywhere. This is because if we have a polynomial like $P(x, y) = 4x^3y + 7xy^2 - 2x + 9$, then

$$\lim_{(x, y)\to(a, b)} P(x, y) = P(a, b)$$

And this is true for every point (a, b) in the plane. More complicated functions constructed from continuous functions by addition, subtraction, multiplication, division, and composition are also continuous at points where they are defined.

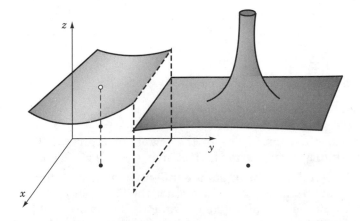

Figure 8.11 A discontinuous function.

Example 1 $f(x, y) = e^{\sin(xy)} + \ln(3x^2 y^3)$ is continuous everywhere it is defined.

This is because it is constructed by stringing together continuous functions like $\sin$, $\cos$, $\ln x$, e^x, and $\sqrt{x}$. They are strung together by $+$, $-$, $\times$, and division and composition. Be careful with division. Division can lead to dividing by zero, which may cause a function not to be defined at some points. We also need to remember that square roots don't like to be fed negative numbers, and $\ln x$ requires that x be positive.

So this function is continuous everywhere it is defined. But where exactly is the function above defined? Well, the first term $e^{\sin(xy)}$ is defined for all x and y since $\sin x$ and e^x are defined everywhere. But the second term $\ln(3x^2 y^3)$ is another story. The function $\ln x$ requires a positive number x to be fed into it, which means $\ln(3x^2 y^3)$ is defined where $3x^2 y^3$ is positive. This is the set of points (x, y), where $x \neq 0$ and $y > 0$. On these points, which are the domain of the function, $f(x, y)$ is continuous.

In general, one should think of continuous functions as friendly and supportive. But discontinuous functions are dangerous and difficult to deal with. They should be avoided if possible, and otherwise treated with care.

Hey, whatever greases your ball bearings.

8.5 Partial derivatives

Remember how the derivative of $f(x)$ gives the slope of the tangent line to the graph of $y = f(x)$, as in Figure 8.12. We would like to extend this idea to functions $f(x, y)$ of two variables.

Okay, so now suppose you find yourself standing on the side of Mount Everest, wearing your incredibly expensive mountaineering equipment, colored neon yellow and black, the same clothes that made you feel so cool when you tried them on at L. L. Bean. But now you're not feeling so cool. Now you

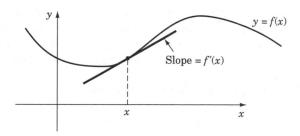

Figure 8.12 The derivative of $f(x)$ gives the slope of the tangent line to the graph of $y = f(x)$.

are downright cold. The wind is whistling up through the bottom of the coat since you forgot to tighten one of the high-tech drawstrings. You are tethered to a rope that's hooked into a piton jammed into some crack in the rock above your head, and the other end of the rope is held by the Finnish expedition leader. He is gesturing to you desperately and yelling Finnish gibberish. It seems like ever since the expedition began, all he has been doing is gesturing at you desperately and screaming Finnish gibberish. The wind is picking up fast and the temperature seems to be dropping at an incredible rate. You find yourself wondering once again why, when you were floating in that pool sipping that tropical drink, you decided you needed more adventure.

You figure the Scandanavian mountain guy is probably saying it's time to head back down to the base camp for a delicious cup of hot cocoa with reconstituted miniature marshmallows, so you take a step in the easterly direction. Unfortunately, in that direction, the next time your foot would touch rock again is 200 feet down in the negative z-direction. Luckily, the piton holds and the Finnish tour guide is now screaming incomprehensibly at you, veins bulging in his forehead, as he grasps the belay rope from which you dangle.

Of course, the problem is that the slope of the mountain in the easterly direction is quite steep. You finally manage to get a foot back on the mountain. You notice that if you step off in the northerly direction, the direction that the Finn is desperately motioning toward, you will touch mountain again only 2 feet down in the negative z-direction. The slope of the mountain in that direction is not so steep.

That gets you to thinking about that multivariable calculus class you had in college. You remember that when you are at a point on a surface in 3-space, as for instance occurs when you are standing on a mountain, there is more than one slope. For any of the compass directions in which you could head, there is a slope, which is the slope of the tangent line in that direction.

Suppose that the surface of the mountain is given by the graph of $z = f(x, y)$, with the positive x-axis pointing in the easterly direction and the positive y-axis pointing in the northerly direction (see Figure 8.13). The fact that as you step east, your foot dangles in the air a long way above the mountain suggests that the tangent line in the x-direction has a large negative slope, say a slope of -200. On the other hand, the tangent line in the y-direction, which is the north direction, although also with negative slope, is not so steep. It has a slope of about -2, meaning that 1 foot out in the y-direction will take you 2 feet down in the z-direction.

You remember how your calculus teacher, the one who never seemed to have been informed about the advantages of personal hygiene (funny what comes back when your life lies in the balance), used to explain that the partial derivative with respect to x, $\dfrac{\partial f}{\partial x}$, was the slope of the tangent line in the x-direction. You remember how you asked him, "Excuse me, Professor

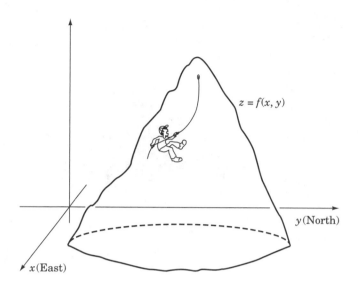

$z = f(x, y)$

$y\,$(North)

$x\,$(East)

Figure 8.13 Climbing a mountain.

Whiner, but there isn't a tangent line in the x-direction, since the x direction is horizontal. All the tangent lines are slanted."

He looked at you with a glare that implied the slime mold on his toothbrush could beat you at tic-tac-toe, and said, "Look, what I mean is you take a plane through your point that is parallel to the xz-plane. And you take the tangent line to the surface that is sitting in that plane. You call the slope of that line $\dfrac{\partial f}{\partial x}(a, b)$." He drew a picture on the board like Figure 8.14.

"Oh," you said, "I see, and then you can compute the derivative in a manner similar to the way you do it in the plane," and you drew a picture like Figure 8.15.

And he actually smiled at you. And even though his two front teeth were horrible to behold, you actually liked him for just an instant, and thought maybe you would be a math major after all.

So we measure the slope of a surface [which is itself the graph of an equation $z = f(x, y)$] in the x-direction. This means we are slicing the surface with a vertical plane that passes through the point $(a, b, f(a, b))$ and that is parallel to the xz-plane, as in Figure 8.14. This plane slices the surface along a curve in the plane. We call the slope of the tangent line to that curve at the point $(a, b, f(a, b))$ the *partial derivative with respect to x of $f(x, y)$ at (a, b),* and we write it as $\dfrac{\partial f}{\partial x}(a, b)$.

Since the vertical plane with which we are slicing turns the three-dimensional situation into a two-dimensional situation, we can find the

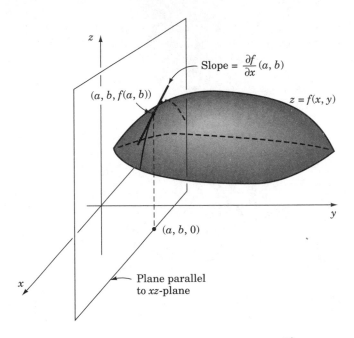

Figure 8.14 The slope of this tangent line is given by $\dfrac{\partial f}{\partial x}$.

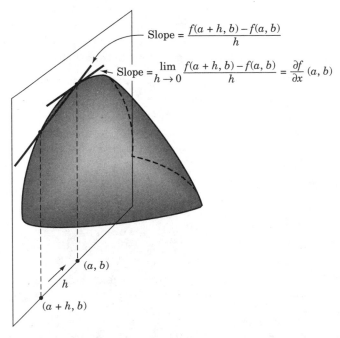

Figure 8.15 Computing the partial derivative $\dfrac{\partial f}{\partial x}(a, b)$.

slope of the tangent line the same way we did way back in single-variable calculus, by taking another point on the curve, figuring out the slope of the resulting secant line, and taking the limit as the second point approaches the first, as in Figure 8.15.

In our case, this gives:

The **partial derivative** of f with respect to x is given by

$$\frac{\partial f}{\partial x} = \lim_{h \to 0} \frac{f(x + h, y) - f(x, y)}{h}$$

Similarly, $\frac{\partial f}{\partial y}(a, b)$ will be the slope of the tangent line to the surface that lies in the plane that passes through the point $(a, b, f(a, b))$ and that is parallel to the yz-plane, as shown in Figure 8.16.

The **partial derivative** of f with respect to y is given by

$$\frac{\partial f}{\partial y} = \lim_{h \to 0} \frac{f(x, y + h) - f(x, y)}{h}$$

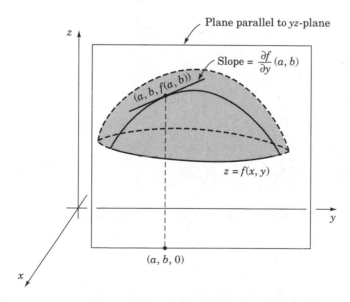

Figure 8.16 The slope of this tangent line is given by $\frac{\partial f}{\partial y}(a, b)$.

Notice that in both cases, all of the action is in one of the two variables. The other variable is just treated as a constant. In fact, that is how we actually evaluate these partial derivatives. We treat the other variable as a constant and differentiate accordingly. Let's make a big deal over that:

To find $\dfrac{\partial f}{\partial x}$, treat y as a constant, and differentiate with respect to x.

So if $f(x, y) = 2x + y$, then $\dfrac{\partial f}{\partial x} = 2$.

We get this by pretending that y isn't a variable at all, but a lowly constant that gets wiped out by differentiation.

Notation   There are a few different ways to write this derivative:

$$f_x = 2$$

or

$$\frac{\partial f}{\partial x} = 2$$

or

$$\partial_x f = 2$$

or

$$D_x f = 2$$

Let's try this on some other examples:

Example 1   If $f(x, y) = x^2 y$, then $\dfrac{\partial f}{\partial x} = 2xy$.

Example 2   If $f(x, y) = \sin x \cos y$, then $\dfrac{\partial f}{\partial x} = \cos x \cos y$.

Important Point, Useful in Later Life (or at least in the next chapter)   In each case, what you get when you take this derivative with respect to x is a new function, and it is a function of exactly the same number of variables as your original function. So to be more careful we could have written:

$$f(x, y) = 2x + y$$

$$\frac{\partial f}{\partial x}(x, y) = 2$$

Now of course, x isn't any more special than y; it just comes first in the alphabet. Those of you whose last names begin with z can appreciate how unfair this favoritism is. So we can differentiate these functions just as well with respect to y, obtaining:

$$f(x, y) = 2x + y$$

$$\frac{\partial f}{\partial y} = 1$$

and

$$f(x, y) = x^2 y$$

$$\frac{\partial f}{\partial y} = x^2$$

and

$$f(x, y) = \sin x \cos y$$

$$\frac{\partial f}{\partial y} = (\sin x)(-\sin y)$$

Now this comes in handy when you're sitting at Thanksgiving dinner and Cousin Ralph is describing in excruciating detail how best to slip shoes on elderly ladies with bunions, and the ratio of sweet potatoes to humans is 10:1, making the ratio of minimarshmallows to humans astronomical. You can excuse yourself from the convivial company to practice taking partial derivatives of partial derivatives, or even partial derivatives of partial derivatives of partial derivatives, like this:

$$f(x, y) = 2x + y$$

$$\frac{\partial f}{\partial y} = 1$$

$$\frac{\partial}{\partial x}\left(\frac{\partial f}{\partial y}\right) = \frac{\partial^2 f}{\partial x \, \partial y} = 0$$

We have taken the partial derivative with respect to x of $\dfrac{\partial f}{\partial y}$. Since there were no x's in the expression for $\dfrac{\partial f}{\partial y}$, we got 0.

The little 2 in the numerator of $\dfrac{\partial^2 f}{\partial x\, \partial y}$ tells you how many partial derivatives to take. The denominator tells you what order to take them in, reading from right to left. This process is like taking a second derivative for functions of one variable, only now we have more than one possibility.

This makes you appreciate some of the other ways of writing this; we could write

$$f(x, y) = 2x + y$$
$$f_y = 1$$
$$f_{yx} = 0$$
$$f_{yxy} = 0$$

This one sort of peters out after a while. How about $f(x, y) = x^2 y$? Here's a random sample of its partial derivatives:

$$f(x, y) = x^2 y$$
$$f_x = 2xy$$
$$f_{xy} = 2x$$
$$f_{xx} = 2y$$
$$f_{xxy} = 2$$

A little more stamina than the linear function, but running out soon.

Now $f(x, y) = \sin x \cos y$ — there's a function that will see you through Thanksgiving and well into New Year's, too:

$$f(x, y) = \sin x \cos y$$
$$f_x = \cos x \cos y$$
$$f_y = -\sin x \sin y$$
$$f_{yx} = -\cos x \sin y$$
$$f_{xx} = -\sin x \cos y$$
$$f_{xxx} = -\cos x \cos y$$
$$f_{xxxx} = \sin x \cos y$$

This function has an endless number of partial derivatives; you can take them one after another, in any order, they never quit. We just stuck on all those x's so that we'll get a lot of hits when we post this on our Web site.

Notice that when we take the so-called *second partials,* we have four possibilities, namely, f_{xx}, f_{yy}, f_{xy}, and f_{yx}. That's at least one too many to keep track of. Luckily the gods of mathematics were kind to us. In their beneficence, they decided to help us out and make $f_{xy} = f_{yx}$ whenever these two are continuous. We call these two derivatives the *mixed second partials.* So what we're saying is that the mixed second partials are equal. Let's highlight this fact. When all these partial derivatives are continuous,

$$\frac{\partial^2 f}{\partial y\, \partial x} = \frac{\partial^2 f}{\partial x\, \partial y}$$

Since pretty much all the functions we ever deal with are continuous wherever defined, as are their derivatives, we don't have to worry too much about that condition.

Example 3   Check that the mixed second partials of $f(x, y) = x^2y^3 + 7y \sin x + 6x$ are equal.

Solution

$$f_x = 2xy^3 + 7y \cos x + 6$$

so

$$f_{xy} = 6xy^2 + 7 \cos x$$

On the other hand,

$$f_y = x^2 3y^2 + 7 \sin x$$

so

$$f_{yx} = 6xy^2 + 7 \cos x$$

Lo and behold, $f_{xy} = f_{yx}$, another miracle brought to you by our sponsors.

SO WHAT ARE PARTIALS GOOD FOR?

We are so glad you asked that question. Let's talk about mountains for a second. Suppose we want to climb Mount McDinkley, the highest mountain in Parmesan County. And suppose that the mountain is given by the graph of $z = f(x, y)$. In order to climb to the top of the mountain, we need to be

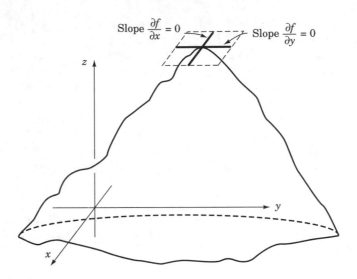

Slope $\dfrac{\partial f}{\partial x} = 0$ Slope $\dfrac{\partial f}{\partial y} = 0$

Figure 8.17 The tangent lines are horizontal at a peak.

able to recognize when we are at the top. How do we tell when we are at the peak? At that point, any tangent line is horizontal, meaning its slope is 0. But $\dfrac{\partial f}{\partial x}$ and $\dfrac{\partial f}{\partial y}$ are slopes of two particular tangent lines, so at the peak (as in Figure 8.17), it must be the case that

$$\frac{\partial f}{\partial x} = 0$$

$$\frac{\partial f}{\partial y} = 0$$

We can use this idea to find peaks of mountains. We just take the two partial derivatives, set them equal to 0 and then solve to find points (x, y) where they are both 0. These points are candidate points (called *critical points*) for where the peak can occur.

In fact, this same idea works to find valleys, too. And we don't have to limit ourselves to Mount McDinkley, or any other mountain for that matter. If $f(x, y)$ is any old function of two variables, be it a profit function, or the number of bacteria in the tupperware in the fridge, this technique will allow us to find the maximum or minimum of the function. We will do exactly that in the next section.

Hey, whatever starches your toga.

8.6 Max-min problems

In the real world we hold so dear, the most common problem we confront is to maximize or minimize some quantity subject to some constraint. For instance, maybe we're on a diet and want to maximize the number of desserts we eat while keeping our intake below 125 calories. Or we have 1 week to study for four finals, and we want to apportion that study time to maximize our resultant grade point average. Or perhaps we have $2.74 and we want to maximize our caffeine buzz with only that much to spend.

We call such a problem a *constrained max-min problem*. Let's try a simple example.

Example 1   Find three numbers x, y, and z that add up to 120 and such that their product is a maximum.

Solution   This is a classic type of problem, dating back to the Stone Age. But in the Stone Age, the biggest known number was 7, so the problems were easier.

Step 1 Set up the function to be maximized and the so-called *constraint equation*. Let $f(x, y, z) = xyz$. This is the function we want to maximize. We are constrained by the equation $x + y + z = 120$.

Step 2 Use the constraint equation to reduce the number of variables to two. Since we can solve the constraint to obtain $z = 120 - x - y$, we can substitute that expression in for z:

$$f(x, y) = xy(120 - x - y) = 120xy - x^2y - xy^2$$

Now we just have a function of two variables for which we want to find the maximum and minimum.

Step 3 Take the partial derivatives and set them equal to 0.

$$\frac{\partial f}{\partial x} = 120y - 2xy - y^2$$
$$\frac{\partial f}{\partial y} = 120x - x^2 - 2xy$$

Setting each of these equal to zero, we get the pair of equations:

$$120y - 2xy - y^2 = 0$$
$$120x - x^2 - 2xy = 0$$

Step 4 Solve these equations for the critical points.
From the first we can factor out a y:

$$y(120 - 2x - y) = 0$$

From the second we can factor out an x:

$$x(120 - x - 2y) = 0$$

If x or y equaled 0, we would get a minimum rather than a maximum. So we can assume they are not 0. Then

$$120 - 2x - y = 0$$
$$120 - x - 2y = 0$$

Subtracting the second equation from the first, we have

$$-x + y = 0$$
$$x = y$$

Substituting this back into the first equation gives

$$120 - 2x - x = 0$$
$$3x = 120$$
$$x = 40$$

It then follows from plugging $x = 40$ into the equation $x = y$ which we saw above that $y = 40$ also, and since $z = 120 - x - y$, we have that $z = 40$ as well. So we should choose all three of our numbers to be 40, in which case the product is $(40)^3 = 64,000$.
Wow! No small potatoes.

Let's try another one.

Example 2   Cleopatra wants a box made to hold her jewels. It is to be 8 cubic cubits. The bottom is to be made of gold sheet, the four sides to be made of silver sheet, and the top is to be made of pearl-encrusted sheet. She offers you 1000 shekels to make the box. If gold sheet costs 16 shekels per square cubit, silver sheet costs 6 shekels per square cubit, and pearl-encrusted sheet costs 8 shekels per square cubit, should you take the commission?

Solution   Let the dimensions of the box be x cubits long, y cubits deep, and z cubits high. How big is a cubit? Who cares? It's actually irrelevant to the problem. (But since you asked so nicely, this ancient measure is the length from the elbow to the tip of the extended middle finger, about 18 inches. Ancient peoples always extended their middle fingers. Nowadays, people only do that some of the time. The Egyptian cubit added to this the width of the

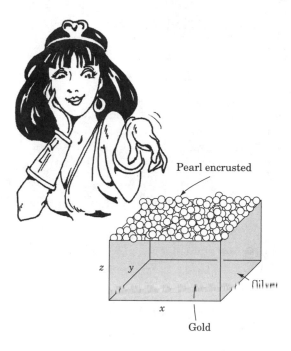

Figure 8.18 A jewel box for Cleo.

palm of the hand of the ruling Pharaoh, giving about 21 inches depending on the Pharaoh. Good enough to build the pyramids.)

Step 1 Draw a picture as in Figure 8.18. Throw in Cleo the Queen of Egypt for good measure.

Step 2 Find a "constraint" on the variables.

  The fact that the box has to hold 8 cubic cubits gives us a relation between the three variables, namely that $xyz = 8$. This is called the constraint equation, since it puts a constraint on the three variables. No more free-for-all. Put a leash on two of them and the third is on a leash, too.

Step 3 Determine the function to be maximized or minimized.

  Our goal is to minimize the cost of making the box. So let's figure out that cost. The cost of the top is the area of the top in square cubits times the cost per square cubit.

$$\text{Cost of top} = (\text{area of top}) \times (\text{cost of top per square cubit}) = (xy)(8)$$
$$\text{Cost of four sides} = (\text{area of four sides}) \times (\text{cost of sides per square cubit})$$
$$= (xz + xz + yz + yz)(6) = 12xz + 12yz$$
$$\text{Cost of bottom} = (\text{area of bottom}) \times (\text{cost of bottom per square cubit}) = (xy)(16)$$

The total cost is

$$C(x, y, z) = \text{(cost of top)} + \text{(cost of sides)} + \text{(cost of bottom)}$$
$$= 8xy + (12xz + 12yz) + 16xy = 24xy + 12xz + 12yz$$

Step 4 Reduce the function to be maximized or minimized to two variables using the constraint equation.

We will use the constraint equation to reduce the cost function to just two variables. Since $z = 8/(xy)$, we have

$$C(x, y) = 24xy + 12x\frac{8}{xy} + 12y\frac{8}{xy} = 24xy + \frac{96}{y} + \frac{96}{x}$$

Step 5 Take the partial derivatives, set them equal to 0, and solve the resulting equations.

$$\frac{\partial C}{\partial x} = 24y - \frac{96}{x^2}$$
$$\frac{\partial C}{\partial y} = 24x - \frac{96}{y^2}$$

Setting each of these equal to zero to find a minimum, we get

$$24y - \frac{96}{x^2} = 0$$
$$24x - \frac{96}{y^2} = 0$$

We want to solve these two equations simultaneously. Solving the first for y, we have

$$y = \frac{4}{x^2}$$

Plugging that into the second equation gives

$$24x - \frac{96}{(4/x^2)^2} = 0$$
$$24x - 6x^4 = 0$$
$$6x(4 - x^3) = 0$$
$$x = 0 \text{ or } x = 4^{1/3} \approx 1.587 \text{ cubits}$$

The solution $x = 0$ doesn't correspond to a minimum, or even to a finite box. But when $x = 4^{1/3}$, then we find $y = 4^{1/3}$ and $z = 2(4^{1/3})$.

The total cost for the box is

$$C = 24(4^{1/3})(4^{1/3}) + \frac{96}{4^{1/3}} + \frac{96}{4^{1/3}}$$

$$\approx 181.42 \text{ shekels}$$

That means we make a profit of about 818 shekels. Wow! That's enough to buy a new thatched roof for the hovel! Things are looking up!

Example 3 Find the dimensions of the box of maximum volume that has one corner at the origin, three sides that contain that corner lying in the three coordinate planes and the opposite corner lying on the plane $2x + 3y + 4z = 12$ in the first octant.

Solution So first let's ask ourselves. Why do we want to know the answer to this question? It could be that we have an attic with a sloping roof and we want to build the biggest box we can in that attic to hold the deformed monster that is the product of years of inbreeding within our royal family. Of course, that's probably a long shot. More likely we make the deformed monster the new king and pretend there's nothing wrong with him.

Or maybe the volume of the box represents our profit as a function of three variables like the amount of capital we invest, the cost of labor, and the demand for our product. And maybe those three variables are constrained to be related by the equation $2x + 3y + 4z = 12$. Or maybe this problem is on the midterm, and that's reason enough to want to know how to do it. Whatever the reason, let's see what we can do with it.

Step 1 First we draw a picture as in Figure 8.19.

Step 2 Find the constraint equation, and the function to be maximized.

Notice that we have set it up so that if (x, y, z) is the coordinate of that corner that lies on the plane $2x + 3y + 4z = 12$, then x, y, and z are also the dimensions of the box. So the volume of the box is given by $V = xyz$, where x, y, and z are constrained by the equation $2x + 3y + 4z = 12$.

Step 3 Use the constraint to lower the number of variables in the function to be maximized or minimized.

Solving the constraint for x, we have $x = 6 - \frac{3}{2}y - 2z$. Substituting this in, we find:

$$V = (6 - \tfrac{3}{2}y - 2z)yz$$
$$= 6yz - \tfrac{3}{2}y^2z - 2yz^2$$

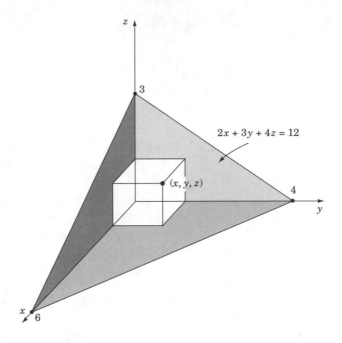

Figure 8.19 A box in the attic.

Step 4 Take the partials

$$\frac{\partial V}{\partial y} = 6z - 3yz - 2z^2$$

$$\frac{\partial V}{\partial z} = 6y - \tfrac{3}{2}y^2 - 4yz$$

and set them equal to 0:

$$6z - 3yz - 2z^2 = 0$$

$$6y - \tfrac{3}{2}y^2 - 4yz = 0$$

The first equation becomes $z(6 - 3y - 2z) = 0$.
The second equation becomes $y(6 - \tfrac{3}{2}y - 4z) = 0$.
The solutions where $z = 0$ or $y = 0$ are giving zero volume and are not maxima, so we can divide by z and y and get

$$6 - 3y - 2z = 0$$

$$6 - \tfrac{3}{2}y - 4z = 0$$

Doubling the first equation and subtracting the second from it gives

$$6 - \tfrac{9}{2}y = 0$$

so $y = \frac{4}{3}$. Then $z = 1$, and from the constraint equation, we find $x = 2$. So the maximum volume is $V = (2)(\frac{4}{3})(1) = \frac{8}{3}$, and the dimensions of the box are $2 \times \frac{4}{3} \times 1$.

It will be a little cramped, but he should get used to it after a while. Hey, whatever restrains your relative.

8.7 The chain rule

Hey, before we get into the chain rule for functions of more than one variable, let's take a leisurely stroll down memory lane with the chain rule of one variable. Remember how this went? There were two ways to describe it. First, let $w = f(u)$ be a function and let $u = g(x)$. So $w = f(g(x))$. That is to say, w depends indirectly on x through u. For instance, we might have $w = \sin u$ where $u = x^2$. So actually, $w = \sin(x^2)$. Then we had:

Chain Rule for Functions of One Variable

$$\frac{d(f(g(x)))}{dx} = f'(g(x))g'(x)$$

The other way the chain rule was often stated was

$$\frac{dw}{dx} = \frac{dw}{du}\frac{du}{dx}$$

Note that these two are the same since $f'(g(x)) = \dfrac{dw}{du}$ and $g'(x) = \dfrac{du}{dx}$. Many people like the second version because it looks like it's canceling the du's.

In our specific example, $\dfrac{d(\sin(x^2))}{dx} = (\cos(x^2))(2x)$.

Now onto:

THE CHAIN RULE FOR FUNCTIONS OF MORE THAN ONE VARIABLE

Suppose we have a function $w = f(u, v)$ where $u = g(x, y)$ and $v = h(x, y)$. So w depends directly on u and v and they each depend directly on x and y. We might depict this particular situation with the diagram shown in Figure 8.20.

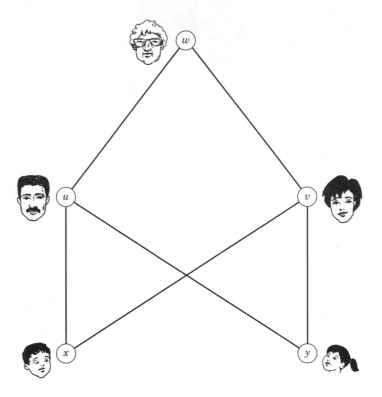

Figure 8.20 How the variables depend on one another.

We call w the *dependent variable,* since it depends directly on u and v and indirectly on x and y. The variables u and v are called the *intermediate variables* and x and y are the so-called *independent variables* as they do not depend on any of the other variables.

We can think of this situation as the chain of financial support that runs through wealthy families. The grandparent w is giving money to the parents u and v, who in turn give the money to the children x and y. So indirectly, the children are being supported by the grandparents.

Now, suppose we want to know the rate at which child x is draining the financial resources of the grandparent w. Then we are asking for $\dfrac{\partial w}{\partial x}$.

The chain rule for functions of more than one variable says:

Chain Rule for Functions of More Than One Variable

$$\frac{\partial w}{\partial x} = \frac{\partial w}{\partial u}\frac{\partial u}{\partial x} + \frac{\partial w}{\partial v}\frac{\partial v}{\partial x}$$

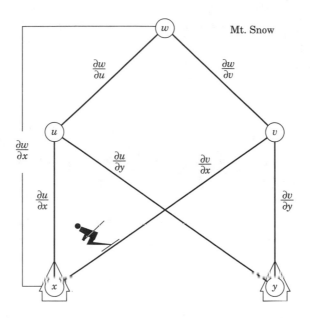

Figure 8.21 Skiing our way through the chain rule.

In other words, the drain on w's money is the drain due to parent u times the drain on u due to x plus the drain due to parent v times the drain on v due to x.

How to remember this? Wealthy people ski a lot and the trick is to treat the diagram as if it is a chart of the ski trails at Mt. Snow. Each edge in the diagram is a trail with name given by the corresponding derivative. So the edge from u down to x corresponds to the derivative $\dfrac{\partial u}{\partial x}$. If we want $\dfrac{\partial w}{\partial x}$, that means we want to get from w at the top of the mountain down to x, which is the ski lodge at the bottom. (See Figure 8.21.)

But we must include each possible trail down. Taking the trail through u, we obtain the term $\dfrac{\partial w}{\partial u}\dfrac{\partial u}{\partial x}$. Taking the trail through v, we obtain the term $\dfrac{\partial w}{\partial v}\dfrac{\partial v}{\partial x}$. Therefore, we have

$$\frac{\partial w}{\partial x} = \frac{\partial w}{\partial u}\frac{\partial u}{\partial x} + \frac{\partial w}{\partial v}\frac{\partial v}{\partial x}$$

Similarly, by including both of the trails from w down to ski lodge y, we find

$$\frac{\partial w}{\partial y} = \frac{\partial w}{\partial u}\frac{\partial u}{\partial y} + \frac{\partial w}{\partial v}\frac{\partial v}{\partial y}$$

Let's try an example to see how we can compute with the chain rule.

Example 1   Let $w = u^2v + 7v - u$. If $u = 2x^2y$ and $v = x + y^2$, find $\dfrac{\partial w}{\partial x}$.

Solution   By our ski slope analysis, we know

$$\frac{\partial w}{\partial x} = \frac{\partial w}{\partial u}\frac{\partial u}{\partial x} + \frac{\partial w}{\partial v}\frac{\partial v}{\partial x}$$
$$= (2uv - 1)(4xy) + (u^2 + 7)(1)$$

Plugging in the expressions for u and v gives an answer entirely in terms of x and y:

$$\frac{\partial w}{\partial x} = (2(2x^2y)(x + y^2) - 1)(4xy) + ((2x^2y)^2 + 7)$$
$$= 20x^4y^2 + 16x^3y^4 - 4xy + 7$$

Notice that this problem could also be done by plugging the expressions for u and v in terms of x and y into the expression for w, to obtain w entirely in terms of x and y and then taking $\dfrac{\partial w}{\partial x}$. But there are many situations when we can't do this. We used the chain rule to get in practice for those situations.

Hey, whatever poaches your salmon.

The gradient and directional derivatives

We're now going to meet this object called the gradient. For now, it's just a cute way to keep track of the two first partials. We put them in a vector, kind of like putting a present in a really nice box to make it look fancy. The finer stores understand this concept. Put a denim shirt in a plastic bag and it's just a denim shirt. But fold it up really nice, with pins, wrap it in tissue, and place it in a black and gold box wrapped with gold ribbon, and suddenly it's a Calvin Klein shirt at three times the price, even if it's the same denim shirt. So we are going to dress up the first partials, by putting them in a pretty vector. No extra charge to you, our reader.

WHAT'S THE GRADIENT?

The *gradient* of a function $f(x, y)$ is given by the formula

$$\nabla f = \left\langle \frac{\partial f}{\partial x}, \frac{\partial f}{\partial y} \right\rangle$$

It's also often written as $\nabla f = \dfrac{\partial f}{\partial x}\mathbf{i} + \dfrac{\partial f}{\partial y}\mathbf{j}$. Seems like a silly way to store information, but as we'll see shortly, it's really handy.

Example 1 Find the gradient of $f(x, y) = x^2y^3 - 3x$.

Solution Well, $\dfrac{\partial f}{\partial x} = 2xy^3 - 3$ and $\dfrac{\partial f}{\partial y} = 3y^2x^2$; so

$$\nabla f = \langle 2xy^3 - 3, 3y^2x^2 \rangle$$

If you're handed a function of three variables $f(x, y, z)$, then the gradient has an extra term:

$$\nabla f = \left\langle \frac{\partial f}{\partial x}, \frac{\partial f}{\partial y}, \frac{\partial f}{\partial z} \right\rangle$$

Hey, whatever plucks your poultry.

DIRECTIONAL DERIVATIVES

We've seen that $\dfrac{\partial f}{\partial x}$ and $\dfrac{\partial f}{\partial y}$ are slopes of tangent lines in the x- and y-directions. But what about all the other tangent lines? They are feeling left out. Maybe we are more interested in the rate of change of the hill in the south by southwest direction instead of the north-south or east-west directions. How do we find these other slopes?

First we pick a unit vector down in the xy-plane, call it $\mathbf{u} = \langle u_1, u_2 \rangle$. This is a bit rude, as we're not supposed to refer to anyone by "Hey $\mathbf{u}$," but after all it's only a vector. Since it's a unit vector, $u_1^2 + u_2^2 = 1$. It points along some compass direction in the plane.

Then we define the rate of change of our function in that direction to be the so-called *directional derivative*, denoted $D_u f$. That is to say, $D_u f$ is the slope of the tangent line in the vertical plane containing $\mathbf{u}$, as in Figure 8.22. But how to find it?

The directional derivative $D_u f$ is given by

$$\boxed{D_u f = \nabla f \cdot \mathbf{u}}$$

If we expand out what this means, it says that

$$D_u f = \nabla f \cdot \mathbf{u} = \left\langle \frac{\partial f}{\partial x}, \frac{\partial f}{\partial y} \right\rangle \cdot \langle u_1, u_2 \rangle = \frac{\partial f}{\partial x}u_1 + \frac{\partial f}{\partial y}u_2$$

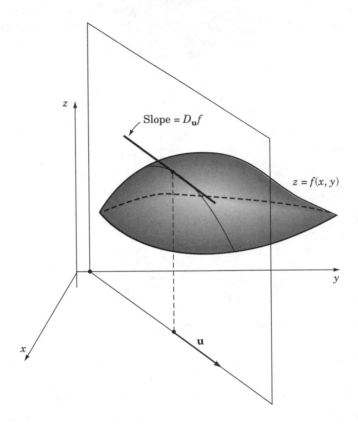

Figure 8.22 The directional derivative is the slope of the tangent line in the vertical plane containing **u**.

Wow, there's that gradient vector again! Popping up all over the place! Amazingly enough, the slope of the tangent line to a surface in any direction at all is given in terms of $\dfrac{\partial f}{\partial x}$ and $\dfrac{\partial f}{\partial y}$. So the other slopes are really all hidden inside the pretty box containing just those two. All the derivative info is in there.

Where did this formula come from? It's obtained by taking the limit

$$\lim_{t \to 0} \frac{f(x + tu_1, y + tu_2) - f(x, y)}{t}$$

So this derivative is defined in terms of a limit just like the derivative of a function of one variable. But we almost never need to use the limit formula to find partial derivatives, just as we didn't with functions of one variable.

The directional derivative $D_u f$ measures the rate of change of f as we move at unit speed in the direction given by **u**.

Let's try our hand at finding one of these.

Example 2   If we are at the point $(1, 2, 19)$ on the hill whose height is given by the function $f(x, y) = 3x^2 + 4y^2$, find the slope of the hill, or the directional derivative, in the direction of $\mathbf{a} = \langle 3, 4 \rangle$. (All units are in feet.)

Solution   Although $\mathbf{a} = \langle 3, 4 \rangle$ gives us a direction in the xy-plane, it is not a unit vector. So we replace it with

$$\mathbf{u} = \frac{\mathbf{a}}{|\mathbf{a}|} = \frac{\langle 3, 4 \rangle}{\sqrt{3^2 + 4^2}} = \langle 3/5, 4/5 \rangle$$

Meanwhile, we can compute the gradient

$$\nabla f = \left\langle \frac{\partial f}{\partial x}, \frac{\partial f}{\partial y} \right\rangle = \langle 6x, 8y \rangle$$

We only care about the gradient at the point $(1, 2, 19)$, so

$$\nabla f(1, 2) = \langle 6, 16 \rangle$$

Then,

$$D_u f = \nabla f \cdot \mathbf{u} = \langle 6, 16 \rangle \cdot \langle 3/5, 4/5 \rangle = 16\tfrac{2}{5}$$

That hill is very steep in this direction. If you want to step 1 foot forward, you will have to go up over 16 feet. Bring some climbing gear.

Common Mistakes

1. Not making the direction vector a unit vector.

2. Thinking the gradient vector is a three-dimensional vector in this type of problem. It's actually a vector in the xy-plane.

Back to the Mountain   The weather has quickly deteriorated, and you find yourself in a swirling snowstorm, where you can see about 10 inches ahead. The Finnish tour guide has managed to ascend to your position, and as far as you can tell, he is trying to push you off the mountain face. You cling to the rock, as he screams gibberish at you and tries to pry your hands free. Every once in a while you kick at him, while hanging on for dear life. This wasn't really the vacation you had in mind. Finally, the tour guide stops, throws his hands up in disgust, and steps off the mountain. You watch as he slides on his rear down the snow covered rock, and you realize that this is a shortcut to the bottom. He is taking the direction of steepest descent, sliding down the steep face on the newly laid snow. But he went so fast, you're not sure which way is the direction of steepest descent. But you know your calculus, and you remember the fact that this portion of the mountain can be closely approximated by a function of form $f(x, y) = y^3 - 3x^2y$ where you are at point (2, 4, 16). You remember the following amazing fact:

> The direction of steepest ascent is in the direction of ∇f, and the directional derivative has magnitude $|\nabla f|$ in this direction. The direction of steepest descent is in the direction of $-\nabla f$, and the directional derivative is $-|\nabla f|$ in this direction.

In your state of confusion, you decide you had better dredge up the reason that this fact is true just to check that you don't have it backward.

Why? The goal is to find the direction for the unit vector $\mathbf{u}$ that will maximize the directional derivative $D_u f$, since when the directional derivative is as large as possible, the corresponding tangent line is as steeply sloped as possible. But the directional derivative is given by

$$D_u f = \nabla f \cdot \mathbf{u} = |\nabla f||\mathbf{u}| \cos \theta$$

where θ is the angle between $\mathbf{u}$ and ∇f. (See Section 5.4.)

Since $\mathbf{u}$ is a unit vector, we know $|\mathbf{u}| = 1$, and therefore we have

$$D_u f = |\nabla f| \cos \theta$$

We want to choose $\mathbf{u}$ to maximize this quantity. However, $|\nabla f|$ is unaffected by the choice of $\mathbf{u}$. The only part we can influence is the $\cos \theta$. Now as everyone knows, $\cos \theta$ can never be bigger than 1. So this quantity will be maximized when $\cos \theta = 1$, implying $\theta = 0$. But if the angle between $\mathbf{u}$ and ∇f is 0, then they both must be pointing in the same direction. And so, we have shown that the maximum for the directional derivative occurs when we head in the direction of ∇f. In this case, since $\cos \theta = 1$, we have that $D_u f = |\nabla f|$.

Similarly, if we want the minimum of the directional derivative, we want to take $\cos \theta = -1$, so $\theta = \pi$, meaning that $\mathbf{u}$ and ∇f point in exactly opposite directions, or that **u** is in the direction of $-\nabla f$. In this case, since $\cos \theta = -1$, $D_u f = -|\nabla f|$.

Convinced that your memories must be right, since you have a proof in your mind, you then calculate the gradient:

$$\nabla f = \left\langle \frac{\partial f}{\partial x}, \frac{\partial f}{\partial y} \right\rangle = \langle -6xy, 3y^2 - 3x^2 \rangle$$

which at the point $(2, 4, 16)$ gives $\nabla f = \langle -48, 36 \rangle$. This gives the direction of steepest ascent. But you want the direction of steepest descent, which is given by

$$-\nabla f = \langle 48, -36 \rangle$$

You launch yourself off the mountain in that compass direction. There is a split second when you are in the air, and then suddenly you are riding down the mountain on the seat of your high tech pants as snow sprays up between your legs. The ride is actually enjoyable until you go over a rock or two, putting into doubt whether you will ever have children.

This fact that the gradient tells you the direction to head for the steepest ascent holds for functions of more variables as well.

Example 3   Suppose $T(x, y, z) = 30x^2 z - y + yz^3$ degrees represents temperature at a point (x, y, z) in the nuclear reactor that, due to the Pepsi you spilled on the control panel, is on the verge of a meltdown. You are in your hotsuit at coordinates $(1, 1, 2)$ feet, having just stepped into the containment chamber, with the instructions to manually turn on the safety water valve. It is uncomfortably hot in there. Determine which direction will make it hotter the quickest.

Solution   If you take ∇T at the point where you are, it will tell you the direction of greatest increase in temperature. In this example,

$$\nabla T = \left\langle \frac{\partial T}{\partial x}, \frac{\partial T}{\partial y}, \frac{\partial T}{\partial z} \right\rangle = \langle 60xz, -1 + z^3, 30x^2 + 3yz^2 \rangle = \langle 120, 7, 42 \rangle$$

So this is the direction you do not want to go. If you move in the direction of this vector, the rate of change of temperature will be $|\nabla T| = \sqrt{120^2 + 7^2 + 42^2} \approx 127°$. So each foot you move in that direction will cause an increase of about $127°$. Don't head that way, if you want to avoid becoming a nuclear french fry.

8.9 Lagrange multipliers

One of the most common types of problems you run up against in the real world is of the type: Maximize or minimize $f(x, y, z)$ subject to a constraint given by $g(x, y, z) = 0$.

Versions of this situation include:

1. Minimize your own weight subject to the constraint that you must eat anything containing chocolate that comes within 10 feet of your mouth.

2. Maximize your salary subject to the constraint that you can't work before 11:00 in the morning or after 3:00 in the afternoon, and a 2-hour lunch break would be nice.

3. Maximize your tan with the constraint that you would like to survive past age 30.

We have already seen one method of solution for maximizing $f(x, y, z)$ subject to the constraint $g(x, y, z) = 0$ in Section 8.6. There, we solved the constraint equation for one of the variables, say z, in terms of the other two. Then we substituted that expression into f to eliminate the z-variable in $f(x, y, z)$. Then we found the critical points of the resulting function of two variables, by taking $f_x = 0$ and $f_y = 0$ and solving them simultaneously.

But all of that only works if we can solve the constraint equation for one of our variables in terms of the other two. What if our constraint equation is

$$x^2y^3 - \sin(xyz) + e^{zy}y + \ln(2x^2 + y + z) = 0$$

Good luck getting any one variable in terms of the others. Not going to happen. Don't bother trying. It's a no go.

So what's a mother to do?

Well, we can tell you what Lagrange's mother did. She turned to little Joseph Louis and said, "Joe Lou, can you come up with a method for solving a problem like this?" She actually said it in French, bu the gist was as above. Now Joe Lou was a momma's boy in the worst way. He would do anything to please his mother, even invent a difficult new mathematical concept. So he went off to his room to work on the problem. When he emerged 20 years later, he had invented the method of Lagrange multipliers. Sadly, while he worked alone in his room for all those years, his mother met a jolly butcher from Lyons named Jacques and she moved away. So when Joey finally emerged, he found himself all alone. It's a sad story, we know, but hey, he got the method named after him, didn't he?

[*Editor's note:* As far as we can tell, all of this is pure fabrication, with the exception that Lagrange did invent the method of Lagrange multipliers. We apologize profusely.]

Here is the method for maximizing or minimizing $f(x, y, z)$ subject to the constraint that $g(x, y, z) = 0$.

Step 1 Make sure the constraint is in the form $g(x, y, z) = 0$. So if our constraint is $x^2 + y^2 = z^2$, we write it as $g(x, y, z) = x^2 + y^2 - z^2 = 0$.

Step 2 Set $\nabla f = \lambda \nabla g$.

Step 3 Solve the resulting set of equations to find x, y, and z.

Step 4 Determine which of the solutions corresponds to the maximum or minimum.

So let's try one of these problems and see how it works.

Example 1   Let the Earth's surface be given by the equation $x^2 + y^2 + z^2 = 1$, with the North Pole at coordinates $(0, 0, 1)$, and Equador at $(1, 0, 0)$. Suppose that an alien species shines a mental enhancement ray in our direction with intensity given by $f(x, y, z) = xy + z$. Where the maximum intensity of the ray occurs on the Earth's surface, we become Einsteins. Where the minimum occurs, we degenerate to the intellectual equivalent of a slug. Find the coordinates on the surface of the Earth where genius will reign.

Solution   It is time to turn to Joe Louie Lagrange. We want to maximize and minimize $f(x, y, z)$ subject to the constraint that $x^2 + y^2 + z^2 = 1$.

Step 1 Put the constraint in the form of $g(x, y, z) = 0$. IT IS VERY IMPORTANT TO DO THIS. HERE IS THE MOST COMMON MISTAKE THAT IS MADE. In our case, we have $g(x, y, z) = x^2 + y^2 + z^2 - 1 = 0$.

Step 2 Set $\nabla f = \lambda \nabla g$. In particular, this means that

$$f_x = \lambda g_x \qquad \text{so } y = \lambda 2x$$
$$f_y = \lambda g_y \qquad \text{so } x = \lambda 2y$$
$$f_z = \lambda g_z \qquad \text{so } 1 = \lambda 2z$$

Step 3 Solve the resulting equations (and the constraint) for x, y, and z. You will probably have to find λ, as well, so go ahead, bite the bullet and find all four. There is no one way to do this. In our example, we will solve the third equation for λ. Since $1 = \lambda 2z$, we know z is not equal to 0, so we can divide by z to get

$$\lambda = \frac{1}{2z}$$

We substitute this into the previous two equations to get rid of λ altogether:

$$y = \frac{x}{z}$$

$$x = \frac{y}{z}$$

Then, plugging the expression for y from the first equation into the second, we have

$$x = \frac{x/z}{z}$$

$$xz^2 = x$$

There are two possibilities. Either $z^2 = 1$ or $x = 0$. But if $x = 0$, then since $y = x/z$, $y = 0$. From our constraint, this would imply $z^2 = 1$. So in any case, $z^2 = 1$.

The end result is that there are two possibilities, $z = 1$ or $z = -1$. In both cases, this forces $x = y = 0$ from the constraint. So, the two critical points are $(0, 0, 1)$ and $(0, 0, -1)$. Plugging in, we see that $f(0, 0, 1) = 1$ and $f(0, 0, -1) = -1$. The first is a maximum for $f(x, y, z)$ and the second is the minimum. It appears that there are going to be some smart polar bears at the North Pole, some very dumb scientists studying the South Pole, and the rest of us will continue to watch professional wrestling.

SO WHAT IS HARD ABOUT LAGRANGE MULTIPLIER PROBLEMS?

You would think that the biggest problem is setting them up, but in fact the setup is usually straightforward. The hard part comes when you have to solve a set of four equations in four unknowns x, y, z, and λ. How do you do that?

There is no single method that is the easiest. But there are a few things to try.

Method 1 Eliminate variables, one at a time. So we may solve one of the equations to get x in terms of y, z, and λ. Then we use that equation to eliminate x from all of the equations. We then repeat the process again and again until we are down to one equation with one unknown which we can then solve.

Method 2 Use the equations to obtain each variable x, y, and z in terms of λ. Then substitute these expressions into the constraint equation to obtain a single equation, all in terms of λ. Then solve for λ, which will then give you the other variables as well.

Common Mistakes in Solving the Equations

1. Oftentimes, when someone is given two equations of the form $h(x) = 0$ and $k(x) - 0$, he or she will set $h(x) = k(x)$ and try to work from there. But this is a huge loss of information. If you say, "The cow is purple," and you say, "The horse is purple," that is a lot more information than just saying, "The cow is the same color as the horse." Same here.

2. Given an equation of the form $x = \lambda x$, many students will cancel the x on each side and say that therefore, $\lambda = 1$. But in fact, canceling the x is equivalent to dividing both sides by x and we can only do that when $x \neq 0$. So in fact there are two possibilities here. Either $\lambda = 1$ or $x = 0$. Either one could lead to the valid solution.

And that is the story of Lagrange multipliers. Hey, whatever ages your Camembert.

8.10 Second Derivative Test

You will remember the famous second derivative test for maxima and minima of functions of one variable.

Second Derivative Test for $f(x)$ If a is a critical point for $f(x)$, so that $f'(a) = 0$, then:

1. If $f''(a) > 0$, then $f(a)$ is a local minimum value.

2. If $f''(a) < 0$, then $f(a)$ is a local maximum value.

3. If $f''(a) = 0$, then the second derivative is telling us a whole lot of nothing.

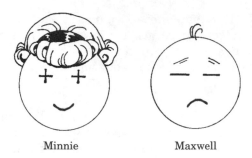

Minnie   Maxwell

Figure 8.23 Minnie and Maxwell tell us whether we have a local min or a local max.

This entire test could be boiled down to two little pictures of Minnie and Maxwell, as in Figure 8.23. Minnie's and Maxwell's eyes give the sign of $f''(a)$. If the eyes are positive signs, the person is happy, with a big smile, which means the lips have a local minimum. If the eyes are negative signs, the person is sad, with a big frown, which gives a local maximum.

But now what happens when we are looking at a function of two variables? Suppose we have a critical point (a, b) in the plane where $\dfrac{\partial f}{\partial x} = 0$ and $\dfrac{\partial f}{\partial y} = 0$. We would like to know whether or not there is a local maximum or minimum at (a, b).

First we will define the so-called discriminant in terms of the second partials.

> The **discriminant** of f is given by $\Delta = f_{xx}f_{yy} - (f_{xy})^2$.

Here is the analog of the second derivative test, one dimension up.

> **The Second Derivative Test for** $f(x)$ If (a, b) is a critical point of $f(x, y)$, then:
>
> **1.** If $\Delta > 0$ and $f_{xx} > 0$, then f has a local minimum at (a, b).
>
> **2.** If $\Delta > 0$ and $f_{xx} < 0$, then f has a local maximum at (a, b).
>
> **3.** If $\Delta < 0$, then f has neither a local max nor min at (a, b) but rather has a saddle point there.
>
> **4.** If $\Delta = 0$, then the second derivative test tells us absolutely nothing.

Minnie   Maxwell   Confused pig   Insane clown

Figure 8.24 Remembering the second derivative test for $f(x, y)$.

There are two things to notice here. First of all, it appears a little strange that f_{xx} seems to play a role here and not f_{yy}. But in fact, in parts 1 and 2, we could have used f_{yy} instead of f_{xx} as they will both have the same sign when $\Delta > 0$. Second, we didn't consider what happens when $\Delta > 0$ and $f_{xx} = 0$. But it's pretty easy to see that given the definition of Δ, this can never occur.

Of course, you may be thinking to yourself, how in heaven am I going to remember these four possiblities? Sometimes one thing is positive, sometimes another. Well, you are in luck, because we have a set of faces that go with these possibilities as well. They appear in Figure 8.24 and go by the names Minnie, Maxwell, the confused pig, and the insane clown.

In each case, the nose represents the sign of the discriminant Δ, and the eyes represent the sign of f_{xx}. The first two resemble Minnie and Maxwell from the second derivative test for functions of one variable. The third is the confused pig, with a negative nose (Δ), and therefore a saddle-shaped mouth. The fourth is the insane clown, with a bulbous nose ($\Delta = 0$) and no information from the mouth.

Okay, let's try an example.

Example 1   Find and classify all the critical points of the function

$$f(x, y) = 2x^3 + y^3 - 3x^2 - 12x - 3y$$

Solution   First we find the first partials:

$$f_x = 6x^2 - 6x - 12 = 0$$
$$f_y = 3y^2 - 3 = 0$$

Then we set them equal to 0:

$$6(x - 2)(x + 1) = 0$$
$$3(y - 1)(y + 1) = 0$$

So $x = 2$ or $x = -1$ and $y = 1$ or $y = -1$.

We obtain four critical points: $(-1, -1), (-1, 1), (2, -1), (2, 1)$.
Taking the second partials, we have

$$f_{xx} = 12x - 6$$
$$f_{yy} = 6y$$
$$f_{xy} = 0$$

So

$$\Delta = (12x - 6)(6y) - (0)^2 = 36y(2x - 1)$$

The following table summarizes everything we need to know at each of the critical points:

Critical point	$f_{xx} = 12x - 6$	$\Delta = 36y(2x - 1)$	Type of point
$(-1, -1)$	-18	108	Local maximum
$(-1, 1)$	-18	-108	Saddle point
$(2, -1)$	18	-108	Saddle point
$(2, 1)$	18	108	Local minimum

Hey, whatever pins the tail on your donkey.

Multiple Integrals

It was looking dicey for Barney Flabbin. "I will establish conclusively," said the student judicial affairs prosecutor, "that this student cheated. In his midterm he performed flawlessly, calculating double and triple integrals in rectangular, polar, cylindrical, and spherical coordinates. He even provided an elegant discussion of the official definition of the double integral in terms of Riemann sums, and that wasn't a question on the exam. Yet when asked afterward, he was unable to do a simple double integral over a rectangle, and in fact he can't now find the center of mass of a solid hemisphere. The only explanation is that he cheated."

The presiding faculty member turned to Barney. "Mr. Flabbin, have you anything to say in your defense?"

"Yes," said Barney slowly. The charges were serious, and could lead to expulsion. The hearing room was so quiet you could hear an epsilon drop. "It's really not my fault. You see, I didn't actually take the test. It was my alternate personality, Fred. He knows about double and triple integrals, but I don't."

There was pandemonium in the courtroom. Barney was claiming multiple integral disorder.

In order to make sure this never happens to you, let's learn about multiple integrals.

Just what are multiple integrals? In brief, a multiple integral is a limit of a Riemann sum over a region in the plane or in space. Huh, what? Maybe that's a little too brief.

Let's think back to what integrals really are. In a word, they're about multiplication. We use an integral to calculate area (width × height), work (force × distance), revenue (units sold × price), etc. Now we learned how to multiply two numbers by third grade, but with integration we can multiply more interesting things. With integrals we can calculate area (width × height) not just when the height is constant, but also when the height is varying from point to point. We can calculate work done (force × distance) not just when the force is constant and we move in a straight line, but also when the force is varying from point to point and the path is bending every which way. We do this by breaking everything up into tiny pieces, so that over each piece the height or force or direction of motion is close to constant. Then we can do our multiplication for each piece just as in third grade, and then add everything up. Finally, we take a limit to see what happens as the pieces get really tiny, and this turns the sum into an integral. We've seen this done for functions of one variable $f(x)$, but it works just as well with functions of several variables $f(x, y)$ or $f(x, y, z)$, giving rise to multiple integrals.

Going from calculating area to calculating volume is the best illustration of how it all works. To get the area of a rectangle, we multiply its width times its height. To find the area under a curve $y = f(x)$ between $x = a$ and $x = b$, we saw in first-semester calculus that we can break $[a, b]$ into lots of small intervals of length Δx_i. Then we can form a rectangle over each interval with height $f(x_i)$ where x_i is a point in the ith interval. The area of each rectangle is $f(x_i) \cdot \Delta x_i$. Adding up these areas gives an approximation to the area under the curve, as in Figure 9.1a. When we take the limit as the number of intervals goes to infinity, we get an integral that gives the total area exactly.

We can use the same ideas to calculate volumes. Just as area is a limit of the areas of a bunch of thin rectangles, so volume is a limit of the sum of the volumes of a bunch of thin boxes. The volume of a box is given by area of base × height. To calculate the volume under $z = f(x, y)$ over a region R in the xy-plane, we break the region R into lots of tiny rectangles of size Δx by Δy. The area of each little rectangle is $\Delta x \cdot \Delta y$, or ΔA for short. We form a box above each tiny rectangle with base the rectangle and height $f(x, y)$, where (x, y) is some point in the rectangle. The volume of the box is $f(x, y) \cdot \Delta A$. Then we add up all the volumes of the skinny boxes as in Figure 9.1b. If we take a limit as the number of boxes goes to infinity, we get the exact volume under the surface.

The same idea shows up when calculating lots of other quantities, like total mass (volume × density), total rainfall (area × rainfall in that area), etc. The main idea behind multiple integrals is this process of breaking up a

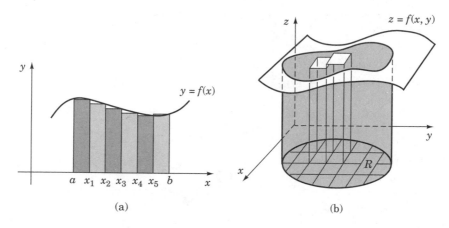

Figure 9.1 Cutting an area into skinny rectangles and a volume into skinny boxes.

region into a bunch of tiny regions where a function is pretty much constant, and multiplying the area of each by the value of the function there.

The other thing we'll need to understand is how to calculate these gadgets. We'll see how to compute a double integral by doing two single integrals of the type we know well, and a triple integral by doing three. Except for a few technicalities, we already know how to calculate multiple integrals.

9.1 Double integrals and limits—the technical stuff

The precise meaning of the double integral is given by a Riemann sum type formula:

$$\iint\limits_{R} f(x, y)\, dA = \lim_{n \to \infty} \sum_{i=1}^{n} f(x_i, y_i)\, \Delta A_i$$

This may look scary, but it's just a repeat of the ideas in the Riemann sum. We break the region R, called the *base*, into a lot of little rectangles of area ΔA_i and in each one we pick a point (x_i, y_i). Each little rectangle will form the base of a rectangular box.

The value of f at (x_i, y_i) will be the height of the corresponding box. The volume of the box is then given by $f(x_i, y_i)\, \Delta A_i$. The sum of all these volumes of slender boxes

$$\sum_{i=1}^{n} f(x_i, y_i)\, \Delta A_i$$

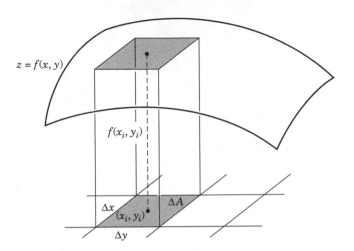

Figure 9.2 A box of height $f(x_i, y_i)$ above each little rectangle.

gives an approximation to the volume we are after. See Figure 9.2. This approximately gives the double integral when the number of little rectangles n is large. Taking the limit as R is divided into more and smaller rectangles, whose number $n \to \infty$, gives the double integral $\iint\limits_R f(x, y)\, dA$.

Technical Point   The limits may not exist if the function $f(x, y)$ is very nasty, or if the region R is exotic. For functions $f(x, y)$ which are continuous, and for functions we are likely to encounter our first time through calculus, this issue won't come up. The interpretation of double integral as a volume holds when the function $f(x, y)$ is positive.

So now we know, technically speaking, what double integrals are. What remains to be seen is

* How to calculate them

* What they're good for

9.2   Calculating double integrals

The method for calculating multiple integrals can be summarized in a word— *iterate*.

The simplest type of region R lies between two vertical lines $x = a$ and $x = b$, and also between two graphs, below $y = g_2(x)$ and above $y = g_1(x)$, as in Figure 9.3a. Such an R is called *vertically simple*. The key to being a vertical simpleton is that over all values of x, the top function never

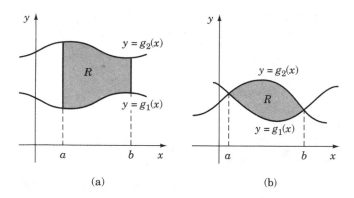

Figure 9.3 Vertically simple regions.

changes, in this case it's always $g_2(x)$, and the bottom function never changes, in this case it's always $g_1(x)$. Another way to put it is that for every vertical line in the xy-plane that intersects the region, as we travel up the line, we hit each curve once, and we always hit the curve $y = g_1(x)$ first and $y = g_2(x)$ second.

Vertical simpletons can also look like Figure 9.3b, where the values of a and b happen to occur where the two graphs intersect.

With this type of region we write the double integral as

$$\int_a^b \int_{g_1(x)}^{g_2(x)} f(x, y)\, dy\, dx$$

It's calculated by doing two integrations, called an *iterated* integral.

$$\int_a^b \left[\int_{g_1(x)}^{g_2(x)} f(x, y)\, dy \right] dx$$

This means that we do the inner integral first:

$$\int_{g_1(x)}^{g_2(x)} f(x, y)\, dy$$

treating x not as a variable, but just as some constant that's hanging around, like a sofa. Once we've done the inner integral, x gets promoted back to a variable. With x a variable we get a function of x, which we then proceed to integrate with respect to x.

We always start with the inner integral and work our way out. A multiple-integral problem, like the problem of losing unsightly fat, is solved by working out.

The limits on the inner integral are called the *inner limits* and the limits on the outer integral are called the *outer limits*.

Example 1 Compute $\int_1^2 \int_x^{x^2} (4x + 10y)\, dy\, dx$.

Solution The region in the xy-plane that we are integrating over is bounded below by $y = x$ and above by $y = x^2$, to the left by $x = 1$ and to the right by $x = 2$. It appears in Figure 9.4.

We compute the inner integral $\int_x^{x^2} (4x + 10y)\, dy$ first. As far as y is concerned, $4x$ is a constant, and its antiderivative is $4xy$. But nobody around here is calling y a constant, and the antiderivative of $10y$ is $5y^2$. So

$$\int_1^2 \int_x^{x^2} (4x + 10y)\, dy\, dx = \int_1^2 \left[4xy + 5y^2 \right]_{y=x}^{y=x^2} dx$$

(evaluating the inner y-integral)

$$= \int_1^2 [4x(x^2) + 5(x^2)^2] - [4x(x) + 5(x)^2]\, dx$$

(plugging in the limits for y)

$$= \int_1^2 4x^3 + 5x^4 - 9x^2\, dx$$

$$= \left[x^4 + x^5 - 3x^3 \right]_1^2 \qquad \text{(doing the outer } x\text{-integral)}$$

$$= [2^4 + 2^5 - 3(2^3)] - [1^4 + 1^5 - 3(1^3)]$$

$$= 24 - (-1)$$

$$= 25$$

Common Mistake Integrating the two integrals in the wrong order is a common mistake. If you do this, the final answer might not turn out to be a number, but could have x's or y's in it. That's bad. Really bad. Always remember to work from the inside integral out.

Important Factoid The outer limits are always constants. (A factoid is something that is either a fact, or if it isn't, should be.) Why should they be constants? If not, the result of taking the iterated integral would have variables in it since, at the end, we plug in the outer limits. So the answer would not be a number, and a number is almost always what we want.

Sometimes we aren't given explicit values for a and b. Instead, we are only given the upper and lower curves bounding the region R.

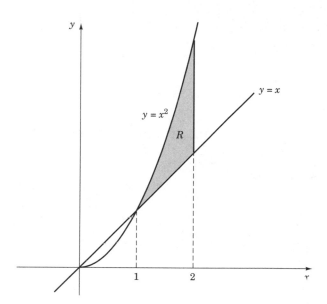

Figure 9.4 The region bounded by $y = x$ and $y = x^2$ between $x = 1$ and $x = 2$.

Example 2   Suppose R is the region between the graphs of $y = x^2$ and $y = 2x$. If $f(x, y) = 8x + 10y$, find $\iint\limits_R f(x, y)\, dx\, dy$.

Solution   This time we aren't given a couple of vertical lines $x = a$ and $x = b$ to cut off the region R. But if we sketch a picture of R as in Figure 9.5 (ALWAYS sketch if you can), then we see how to determine a and b.

The graphs of $y = x^2$ and $y = 2x$ in the xy-plane cross at $(0, 0)$ and at $(2, 4)$. The only finite region bounded by the two curves is the one between $x = 0$ and $x = 2$, so this is our region R. It is a vertical simpleton, so we get the double integral:

$$\int_0^2 \int_{x^2}^{2x} (8x + 10y)\, dy\, dx$$

Working from the inside out, we have

$$\int_0^2 \int_{x^2}^{2x} (8x + 10y)\, dy\, dx = \int_0^2 \left[8xy + 5y^2 \right]_{y=x^2}^{y=2x} dx \quad \text{(doing the inner } y\text{-integral)}$$

$$= \int_0^2 ([16x^2 + 20x^2] - [8x^3 + 5x^4])\, dx$$

$$= \int_0^2 (36x^2 - 8x^3 - 5x^4)\, dx = \left[12x^3 - 2x^4 - x^5 \right]_0^2$$

$$= [12(8) - 2(16) - 32] - [0] = 32$$

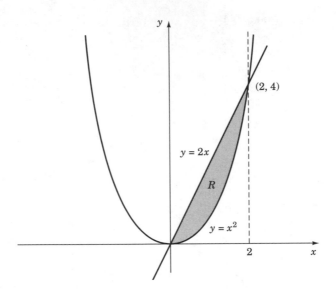

Figure 9.5 The region bounded by $y = x^2$ and $y = 2x$.

For some regions, the roles of x and y are interchanged.

Example 3 Integrate $f(x, y) = y + 1$ over the region R which is bounded by $y = 0, y = 1, x = -y^2 + 4$, and $x = y^2 - 4$.

Solution We first draw the region as in Figure 9.6. The boundary curves of this region have x given as a function of y. Notice that the graph of

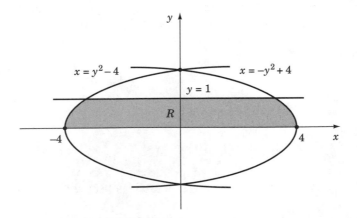

Figure 9.6 The region bounded by $y = 0, y = 1, x = -y^2 + 4$ and $x = y^2 - 4$.

$x = -y^2 + 4$ is always to the right of the graph of $x = y^2 - 4$ when y is between 0 and 1. Put another way, every horizontal line that intersects the region does so to the left along the graph of $x = y^2 - 4$ and then to the right along the graph of $x = -y^2 + 4$. Such a region is called *horizontally simple*. When it comes to humans, a horizontal simpleton is just a vertical simpleton who has gone to bed. But when it comes to integrals, a horizontal simpleton is a region that would be a vertical simpleton, if the roles of x and y were reversed.

When R is a horizontal simpleton, the inner integral becomes the x-integral. In our example:

$$\iint\limits_R f(x, y)\, dA = \int_0^1 \int_{y^2-4}^{-y^2+4} (y + 1)\, dx\, dy$$

Switching the roles of x and y is okay. It's like cutting our pizza into horizontal slices instead of vertical ones. As long as we eat all the slices, we wind up with the same amount of cheese in our tummies. Now it's true that most people don't slice their pizza into long thin pieces, but they might if they had a good set of ginsu knives.

We now compute the integral starting with the inner one and working our way out:

$$\int_0^1 \int_{y^2-4}^{-y^2+4} (y + 1)\, dx\, dy = \int_0^1 \left[yx + x \right]_{x=y^2-4}^{x=-y^2+4} dy \quad \text{(doing the inner } x\text{-integral)}$$

$$= \int_0^1 [y(-y^2 + 4) + (-y^2 + 4)] - [y(y^2 - 4) + (y^2 - 4)]\, dy$$

$$= \int_0^1 8y - 2y^3 - 2y^2 + 8\, dy$$

$$= \left[4y^2 - \frac{y^4}{2} - \frac{2y^3}{3} + 8y \right]_0^1 \quad \text{(doing the outer } y\text{-integral)}$$

$$= 4 - \tfrac{1}{2} - \tfrac{2}{3} + 8$$

$$= 10\ \tfrac{5}{6}$$

The trickier part of calculating double integrals often comes before we actually do any integration. We need to describe the region R correctly, and get the limits of integration set up. This is relatively easy for regions that are vertically or horizontally simple.

But a person who is deep and complicated would never be called a simpleton, either vertical or horizontal. Similarly, a region that is complicated enough may be neither a vertical nor horizontal simpleton. What do we do if we need to compute a double integral over such a region? Turn to the old adage, "divide and conquer." Split the region up into pieces that are either vertical or horizontal simpletons, do the integrals over each, and sum up the result.

Example 4   Find the area of the region R in the first quadrant of the xy-plane bounded by $y = x^2$, $y = x^2/8$, and $y = 1/x$.

Solution   First we draw the region as in Figure 9.7. Notice that there is actually only one region in the first quadrant of the xy-plane (the quarter of the xy-plane that has both x and y positive) with finite area that is bounded by all three of the curves. So this must be our region R. To find the intersection point of $y = x^2$ and $y = 1/x$, we set

$$x^2 = \frac{1}{x}$$

$$x^3 = 1$$

$$x = 1$$

So $(1, 1)$ is the intersection point of these two curves.

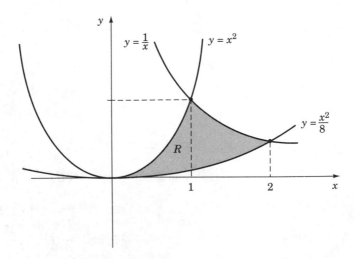

Figure 9.7 The region in the first quadrant bounded by $y = x^2$, $y = x^2/8$, and $y = 1/x$.

To find the point of intersection of $y = x^2/8$ and $y = 1/x$, we set

$$\frac{x^2}{8} = \frac{1}{x}$$

$$x^3 = 8$$

$$x = 2$$

So $(2, \frac{1}{2})$ is the intersection point of these two curves.

Unfortunately, the region is neither vertically simple nor horizontally simple. But if we cut the region R into two pieces, one to the left of $x = 1$ and one to the right of $x = 1$, we have two vertically simple regions. So the total area A will be the sum of the areas of each of these, which is given by integrating the constant function 1. Since $1\,dy\,dx = dy\,dx$, we don't bother writing the "1."

$$A = \int_0^1 \int_{x^2/8}^{x^2} dy\,dx + \int_1^2 \int_{x^2/8}^{1/x} dy\,dx$$

$$= \int_0^1 \Big[y \Big]_{x^2/8}^{x^2} dx + \int_1^2 \Big[y \Big]_{x^2/8}^{1/x} dx$$

$$= \int_0^1 \left(x^2 - \frac{x^2}{8} \right) dx + \int_1^2 \left(\frac{1}{x} - \frac{x^2}{8} \right) dx$$

$$= \left[\frac{x^3}{3} - \frac{x^3}{24} \right]_0^1 + \left[\ln x - \frac{x^3}{24} \right]_1^2$$

$$= \frac{1}{3} - \frac{1}{24} + [\ln (2) - \frac{1}{3}] - [\ln (1) - \frac{1}{24}]$$

$$= \ln 2$$

$$\approx 0.693$$

Hey, whatever toasts your marshmallows.

9.3 Double integrals and volumes under a graph

People always say they're concerned about their weight. "Oh, dear, I've put on 5 pounds." But that's not what they mean. What really concerns them is their volume. They wouldn't care that they weighed 500 pounds if they look trim and fit, with abs of steel. Picture these people—with muscles rippling over bodies that don't show an iota of fatty tissue. Such people probably walk around wearing just the tiniest bikinis, because after all that working out, they're going to strut their stuff. But suppose their bones are made of lead and they weigh in at half a ton. Do they care? Not in the

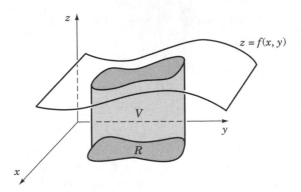

Figure 9.8 The solid V that lies under the graph of $z = f(x, y)$ and above region R in the plane.

least. The only time the issue comes up is when they go horseback riding or step on an elevator with a "weight limit: 450 pounds" sign.

So we would like to be able to calculate volumes of objects, particularly ourselves. Just as a single definite integral allowed us to calculate areas under a graph over an interval $[a, b]$ for a positive function $f(x)$, so the double integral allows us to calculate the volume under a graph of a positive function $f(x, y)$ over a region R. See Figure 9.8.

The volume is obtained by calculating the double integral of the function $f(x, y)$ over the region R.

Example 1 (A Cool Example)   Your brother-in-law Harvey sells speed bumps for a living, and he has an office situated under the freeway overpass. You are supposed to buy an air-conditioner for the office, and you need to know the volume of the room to decide how big a unit to purchase. The cheap unit can cool 20 cubic meters and the expensive one cools 30 cubic meters. Because it was built to fit under the freeway, the room is rather oddly shaped. Taking measurements (in meters), you figure out that the floor is the region R in the xy-plane bounded by $y = x - 1$, $y = x + 1$, $x = 1$, and $x = 2$. The height of the ceiling in meters is given by the function $f(x, y) = 6xy$. What's the volume of the room?

Solution   First let's draw the floor of the room in the xy-plane as in Figure 9.9. Now maybe you're wondering why Harvey would take such an oddly shaped office, but that's not really relevant to the problem. As it happens, he got a really good deal on the rent. What is relevant is that the ceiling is always above the floor, so that $f(x, y)$ isn't negative in our region. If it were, the calculation of volume would first require us knowing where $f(x, y)$ is positive and where it's negative. But in the region R we see that $x \geq 0$ and $y \geq 0$

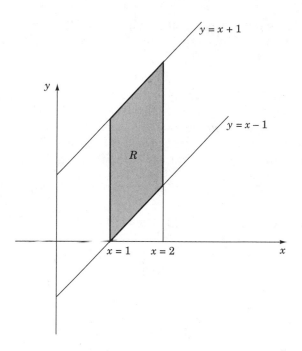

Figure 9.9 The region R that forms the floor of the room.

so that the function $6xy \geq 0$. Since the floor of the room, which is playing the role of the base of our solid (the solid being the room itself) is vertically simple, the volume is given by

$$\text{Volume} = \int_1^2 \int_{x-1}^{x+1} 6xy \, dy \, dx$$

Working from the inside out, we have

$$\int_1^2 \int_{x-1}^{x+1} 6xy \, dy \, dx = \int_1^2 \left[3xy^2 \right]_{y=x-1}^{y=x+1} dx$$

$$= \int_1^2 [3x(x+1)^2] - [3x(x-1)^2] \, dx$$

$$= \int_1^2 [3x(x^2 + 2x + 1)] - [3x(x^2 - 2x + 1)] \, dx$$

$$= \int_1^2 12x^2 \, dx = \left[4x^3 \right]_1^2 = [32] - [4]$$

$$= 28$$

The volume of the room is 28 cubic meters. The cheap air-conditioner will cool only 20, so tell him to spring for the big unit (even though he does live in Alaska).

Common Mistake   Always make sure $f(x, y)$ is positive over the region R when calculating a volume. Just as with areas, if part of the graph dips below the xy-plane, then integration over the corresponding subregion will contribute a negative quantity to the total volume. The integral will no longer represent a volume, but rather the sum of "signed" volumes.

9.4 Double integrals in polar coordinates

When we are dealing with regions bounded by circles, it is often easier to work with polar coordinates. In this setting, points are described by (r, θ) rather than (x, y), and the area of a tiny region is not $dx \cdot dy$ but rather $r\, dr\, d\theta$. The radian measure of angle has the nice property that the length of a piece of a circle of radius r sweeping out an angle of $\Delta\theta$ is $r\,\Delta\theta$. In Figure 9.10, a small wedge in which the radius changes from r to $r + \Delta r$ and the angle changes from θ to $\theta + \Delta\theta$ has area approximately $\Delta r \cdot r\,\Delta\theta$. The extra r-factor comes about because one side of the wedge has length about $r\,\Delta\theta$ rather than just $\Delta\theta$.

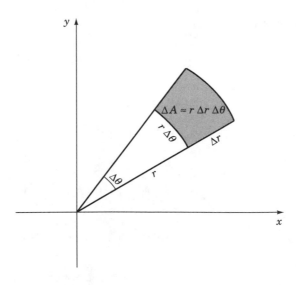

Figure 9.10 The area of a little wedge is approximately $r\,\Delta r\,\Delta\theta$.

So in polar coordinates the double integral becomes

$$\int\int_R f(x, y)\, dA \;=\; \int\int_R f(r, \theta)\, r\, dr\, d\theta$$

> **Important Fact to Remember:** Every double integral in polar coordinates will have that extra factor of r that multiplies $dr\, d\theta$.

Computing double integrals in polar coordinates is done by iterating two single integrals, just as before.

Example 1 Calculate the integral of the function $f(x, y) = x^2 + y^2$ over the region inside the circle of radius 2 and outside the circle of radius 1, centered at the origin.

Solution Polar coordinates really pay off here. Without them we would have a pretty hard time describing the region R, shown in Figure 9.11. In polar coordinates, R is described easily as the (r, θ) for which

$$1 \le r \le 2 \qquad 0 \le \theta \le 2\pi$$

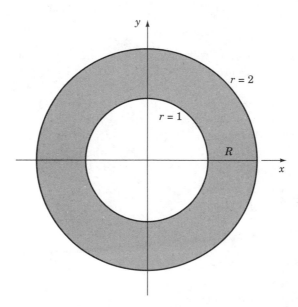

Figure 9.11 The region R between the circles of radius 1 and radius 2.

Since $x^2 + y^2 = r^2$, the function $f(x, y) = x^2 + y^2$ in polar coordinates is $f(r, \theta) = r^2$. Then the double integral becomes

$$\int_0^{2\pi} \int_1^2 r^2\, r\, dr\, d\theta = \int_0^{2\pi} \int_1^2 r^3\, dr\, d\theta$$

$$= \int_0^{2\pi} \left[\frac{r^4}{4} \right]_1^2 d\theta$$

$$= \int_0^{2\pi} {}^{15}\!/_4\, d\theta$$

$$= \left[({}^{15}\!/_4)\theta \right]_0^{2\pi}$$

$$= ({}^{15}\!/_4)(2\pi)$$

$$= ({}^{15}\!/_2)\pi$$

Example 2   You have just started an internet company called Adam's Apple Pies. The orders start flooding in, but you haven't yet figured out how to make an apple pie. If each pie requires a volume of 150 cubic inches of apple, and if the apples have radius 2 inches, how many apples are needed per pie? Keep in mind that the apples must be cored with a corer of radius 1 inch.

Solution   We need to figure out how much volume of apple is left after an apple has been cored. We will assume that each apple is a perfect sphere of radius 2. We will remove from that sphere a vertical cylinder's worth of apple, the cylinder having a radius of 1. (See Figure 9.12.) The surface of the apple is given by a sphere of radius 2, with equation $x^2 + y^2 + z^2 = 4$, or, converting to polar coordinates,

$$r^2 + z^2 = 4$$

Solving for z, we find that

$$z = \pm \sqrt{4 - r^2}$$

The equation $z = \sqrt{4 - r^2}$ gives the top half of the sphere and the equation $z = -\sqrt{4 - r^2}$ gives the bottom half.

    Let's use a double integral to find the volume of the top half of the cored apple and then double the result. The volume that we want will be the solid that is under the surface $z = \sqrt{4 - r^2}$ and above the region in the xy-plane

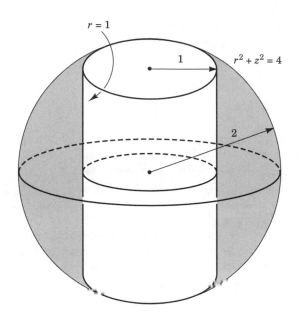

Figure 9.12 Finding the volume of a cored apple.

inside the circle $r = 2$ and outside the circle $r = 1$, so we do not include the core. The total volume of the cored apple is given by

$$\text{Volume} = 2 \int_0^{2\pi} \int_1^2 (\sqrt{4 - r^2}) \, r \, dr \, d\theta$$

The inner r-integral can be done by a u-substitution with $u = 4 - r^2$, and we get

$$\text{Volume} = 2 \int_0^{2\pi} \int_1^2 (\sqrt{4 - r^2}) \, r \, dr \, d\theta$$

$$= 2 \int_0^{2\pi} \left[-\frac{(4 - r^2)^{3/2}}{3} \right]_1^2 d\theta$$

$$= 2 \int_0^{2\pi} [-0] - \left[-\frac{3^{3/2}}{3} \right] d\theta$$

$$= 2 \int_0^{2\pi} \sqrt{3} \, d\theta$$

$$= 2 \left[\sqrt{3}\theta \right]_0^{2\pi} = 4\sqrt{3}\pi$$

$$\approx 21.8 \text{ cubic inches}$$

In order to get a volume of 150 cubic inches of apple per pie, it looks like you need about seven apples per pie. Well, apples are pretty cheap. Maybe you can be the first to make some money with this internet commerce thing.

9.5 Triple integrals

Beverly had developed a taste for integrals as a first-year student. She got intense pleasure from calculating areas, and when she was introduced to integration by parts her ecstasy knew no end. But after a while she stopped feeling the old buzz, and even antiderivatives of inverse hyperbolic functions got stale. She considered studying physics and in a moment of despair, even psychology. But then, a junior math major turned her on to double integrals and the old thrill returned. She was calculating areas, volumes, masses. It was a kick for a while. But soon they, too, were not enough.

She had heard about something stronger, but dangerous. "Don't go near that stuff," her Calc I instructor had warned in his nasal voice. "Once you get into it, you never give it up." But she couldn't help herself. She had to try it.

The house was in the seedy part of town. She saw teaching assistants and postdocs sprawled in doorways amid scraps of paper and old calculus texts. The address had been given to her by a wild-eyed, decrepit senior who, unless he entered treatment soon, would probably be headed for graduate school. Looking furtively over her shoulder, she passed her few remaining dollars through the slot in the door, trembling in anticipation of the final step on her journey: *the triple integral.*

Okay, so maybe integrals aren't mind bending enough to be smuggled in on speedboats, but triple integrals do take the idea of double integrals one step further. To compute an integral over a region in 3-space, we iterate as before, but one extra time.

Triple integrals in *xyz*-coordinates are written like:

$$\int\int\int_V f(x, y, z)\, dV \quad \text{or} \quad \int\int\int_V f(x, y, z)\, dx\, dy\, dz$$

They take their precise meaning from a Riemann sum:

$$\int\int\int_V f(x, y, z)\, dV = \lim_{n \to \infty} \sum_{i=1}^{n} f(x_i, y_i, z_i)\, \Delta V_i$$

where this time V is a solid region in three-dimensional space and the V_i are little boxes with volume ΔV_i.

It's sometimes tricky to draw a picture of the solid V in 3-space. But it is almost always necessary to sketch the solid in order to determine the correct limits of integration.

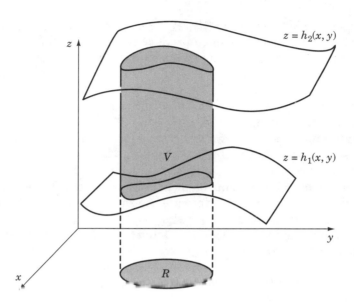

Figure 9.13 A solid V in 3-space bounded by the graphs of two functions over a region R in the xy-plane.

The most basic type of solid V is sandwiched between two surfaces, each a graph over a region in the plane, as in Figure 9.13. The top and bottom boundaries of V are described by two functions, $h_1(x,y) \leq z \leq h_2(x,y)$, and the region in the plane over which this graph sits is described by $g_1(x) \leq y \leq g_2(x)$ and $a \leq x \leq b$. We call such a solid z-*simple* since for any vertical line in 3-space over the region, as we move up the line we first intersect the solid at the surface $z = h_1(x,y)$ and then we leave the solid at the surface $z = h_2(x,y)$.

The triple integral for such a region is calculated iteratively as

$$\int_a^b \left[\int_{g_1(x)}^{g_2(x)} \left[\int_{h_1(x,\,y)}^{h_2(x,\,y)} f(x,y,z)\, dz \right] dy \right] dx$$

The simplest function $f(x, y, z)$ that we might put inside a triple integral is the constant function $f(x, y, z) = 1$. Integrating this function over a region in 3-space gives the volume of that region, and this is one application of the triple integral.

$$\iiint_V 1 \; dx\, dy\, dz = \text{volume } (V)$$

Example 1 (Chicken Not So Little) The famous artist and healer Wynodia has obtained a commission to produce a bronze sculpture for the National Chicken Association. She envisions an abstract representation of a chicken cast in bronze, shaped like the solid above the xy-plane and beneath the paraboloid $z = 9 - x^2 - y^2$. The postmodern chicken would be placed along the road, right in front of the association's headquarters. The title of the work will be "Why did...?" The National Chicken Association has budgeted enough funds to buy 130 cubic meters of bronze. Calculate the volume of the planned sculpture to see if it's too big for its budget.

Solution Even if your major is art, you need to know calculus. And art majors have an easier time drawing the solids besides. (See Figure 9.14.)

We see from the figure that the solid is z-simple. It's pretty clear from the picture that the two surfaces $z = 0$ and $z = 9 - x^2 - y^2$ intersect in a circle. We calculate the radius of this circle by solving $9 - x^2 + y^2 = 0$, which gives $x^2 + y^2 = 9$, a circle of radius 3.

The region in the xy-plane inside this circle forms the base of our solid. Solving $x^2 + y^2 = 9$ for y, we find

$$y = \pm \sqrt{9 - x^2}$$

So the region inside the circle of radius 3 is described by

$$-\sqrt{9 - x^2} \leq y \leq \sqrt{9 - x^2}$$

and

$$-3 \leq x \leq 3$$

The sculpture lies below $z = 9 - x^2 - y^2$ and above $z = 0$. We're calculating volume so we integrate $f(x, y, z) = 1$ over the chicken solid. The triple integral that gives the volume is

$$V = \int_{-3}^{3} \int_{-\sqrt{9-x^2}}^{\sqrt{9-x^2}} \int_{0}^{9-x^2-y^2} 1 \, dz \, dy \, dx$$

So we've described the solid and set up the integral. Now we need to calculate the integral. This process will be essentially the same as for double integrals. We will always start with the innermost integral, which in this case is the z-integral. We will integrate that, treating x and y as constants.

$$\int_{-3}^{3} \int_{-\sqrt{9-x^2}}^{\sqrt{9-x^2}} \int_{0}^{9-x^2+y^2} 1 \, dz \, dy \, dx = \int_{-3}^{3} \int_{-\sqrt{9-x^2}}^{\sqrt{9-x^2}} [z] \Big|_{0}^{9-x^2+y^2} dy \, dx$$

$$= \int_{-3}^{3} \int_{-\sqrt{9-x^2}}^{\sqrt{9-x^2}} 9 - x^2 - y^2 \, dy \, dx$$

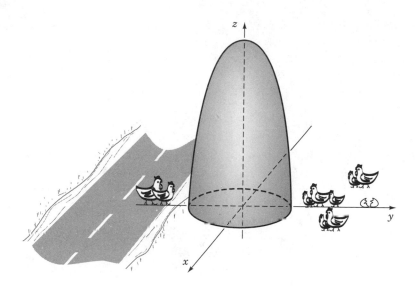

Figure 9.14 The postmodern chicken above the plane $z = 0$ and beneath $z = 9 - x^2 - y^2$.

This double integral over the circular region of radius 3 is easiest if we change to polar coordinates. The planar region that we integrate over is $0 \leq r \leq 3, 0 \leq \theta \leq 2\pi$. The function $9 - x^2 - y^2$ becomes $9 - r^2$, and $dy\, dx$ becomes $r\, dr\, d\theta$. We get

$$\int_0^{2\pi} \int_0^3 (9 - r^2)\, r\, dr\, d\theta = \int_0^{2\pi} \int_0^3 (9r - r^3)\, d\theta$$

$$= \int_0^{2\pi} \left[\frac{9r^2}{2} - \frac{r^4}{4} \right]_0^3 d\theta$$

$$= \int_0^{2\pi} 81/4\, d\theta$$

$$= 81/4(2\pi)$$

$$= 81/2(\pi)$$

$$\approx 127.2$$

Looks like the sculpture's a go.

Note This problem could have been done using a double integral with the function $9 - x^2 - y^2$ inside the integral, instead of using a triple integral. This is because the function we integrated, $f(x, y, z) = 1$, was constant. When we calculate things like center of mass, which we'll see shortly, the function inside the integral won't necessarily be constant, and we will be forced to use a triple integral.

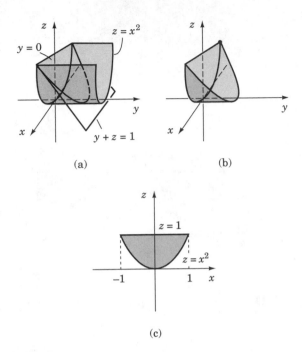

Figure 9.15 A wedge of Gouda cheese.

Example 2 Find the volume of the wedge of Gouda cheese bounded by the three surfaces $y = 0$, $z = x^2$, and $y + z = 1$.

Solution First, we graph the three surfaces. The equation $y = 0$ is just the xz-plane. The equation $z = x^2$ gives a raingutter sitting on the y-axis. And $y + z = 1$ is a slanted plane hitting the y- and z-axes one unit out. (See Figure 9.15a.)

The only solid that is finite in extent and that is bounded by all three surfaces is the wedge in Figure 9.15b. Although this solid is z-simple, it is also y-simple, and it will be slightly easier to use the fact it's y-simple to set up the triple integral. The "base" will be easier to see.

Notice that when we project the solid to the xz-plane, the shadow we see looks like a heel mark (Figure 9.15c) bounded by $z = x^2$ and $z = 1$. This region in the xz-plane is vertically simple, and is described by $-1 \le x \le 1$ and $x^2 \le z \le 1$. For any value of (x, z) that is in this heel-shaped region, the y-values in the solid always satisfy $0 \le y \le 1 - z$. So the volume of the cheese is given by

$$V = \int_{-1}^{1} \int_{x^2}^{1} \int_{0}^{1-z} \, dy \, dz \, dx$$

Since we are using the fact the solid is y-simple, the y-integral goes on the inside. Now, we evaluate the integral from the inside out:

$$\int_{-1}^{1}\int_{x^2}^{1}\int_{0}^{1-z} dy\, dz\, dx = \int_{-1}^{1}\int_{x^2}^{1}\Big[y\Big]_{0}^{1-z} dz\, dx$$

$$= \int_{-1}^{1}\int_{x^2}^{1} 1 - z\, dz\, dx$$

$$= \int_{-1}^{1}\left[z - \frac{z^2}{2}\right]_{x^2}^{1} dx$$

$$= \int_{-1}^{1}\left[1 - \frac{1}{2}\right] - \left[x^2 - \frac{x^4}{2}\right] dx$$

$$= \int_{-1}^{1} \frac{1}{2} - x^2 + \frac{x^4}{2}\, dx = \left[\frac{x}{2} - \frac{x^3}{3} + \frac{x^5}{10}\right]_{-1}^{1}$$

$$= \left[\frac{1}{2} - \frac{1}{3} + \frac{1}{10}\right] - \left[\frac{-1}{2} - \frac{(-1)^3}{3} + \frac{(-1)^5}{10}\right]$$

$$= 1 - \tfrac{2}{3} + \tfrac{2}{10}$$

$$= \tfrac{8}{15}$$

Hey, whatever saddles your pony.

9.6 Cylindrical and spherical coordinates

There are different ways to explain how to find Orzo Quinko, the trendy new restaurant where you've invited your friend for lunch. You could say, "Meet me at 40 degrees 45 minutes latitude, 74 degrees 13 minutes longitude." Most people don't respond well to those directions, with some exceptions in the sailor community.

Or you could say, "Go down Loblolly Street until you see the Citgo Station, take a right at the third light after that, not including the two blinking lights, then at the second oak tree veer to the left (but not all the way to the left) onto Shrinkwrap Way, and the second restaurant on the left is the place." Both describe the same destination, but in different coordinates.

Similarly, we can have more than one description of a point in 3-space. We're already very comfortable with rectangular (or Cartesian) coordinates. But just as polar coordinates were sometimes just the thing to solve two-dimensional problems, there are times when it pays to look at alternative coordinate systems in 3-space.

You may say, "Now why is this the time? I'm very happy working with the rectangular coordinate system. Why make my life difficult?" If you did say that, shame! We're hurt.

Hey, we've stayed up late worrying that maybe some topic could be rewritten just for you, spent long hours thinking about how to make things clear. Cylindrical and spherical coordinate systems will make your life *easier.* At certain critical moments you'll be so glad to see them, you would kiss their feet in gratitude—if they had any.

CYLINDRICAL COORDINATES

Cylindrical coordinates are really easy. All we do is use polar coordinates in the plane and add a regular old z-coordinate. In cylindrical coordinates, a point is identified by (r, θ, z), where r and θ tell you where to go in the xy-plane and the z-coordinate tells how far to go above or below that point. So, for instance, $(r, \theta, z) = (2, \pi/4, 3)$ is the point that appears in Figure 9.16.

The equations to switch between rectangular and cylindrical coordinates are just like the ones for polar coordinates:

$$x = r \cos \theta$$
$$y = r \sin \theta$$
$$r^2 = x^2 + y^2$$
$$z = z$$

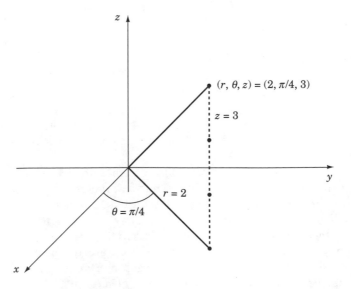

Figure 9.16 The point $(r, \theta, z) = (2, \pi/4, 3)$.

The variable z doesn't really need an equation, since in rectangular and cylindrical coordinates it's the same z. But we put it in anyway so it wouldn't feel left out.

Certain shapes are easy to describe in cylindrical coordinates, specifically cylinders, but also cones, paraboloids, and other regions which are described by their distance from a line. If you're calculating with electric fields around wires, or oil flowing through pipes, or beer swirling around mugs, you're working with cylinders and cylindrical coordinates will likely make your life easier.

Example 1 Describe the surface given in cylindrical coordinates by $r = 3$.

The equation $r = 3$ describes an infinite vertical cylinder of radius 3, encircling the z-axis. How can we see that?

First we see what it looks like in the xy-plane, when $z = 0$. Then $r = 3$ describes a circle of radius 3 in the plane. Since z doesn't appear in the equation, z can be anything. So we take all the points with the same r and θ as those we drew in the xy-plane, but with z-coordinate anything. This gives the infinite vertical cylinder, as in Figure 9.17.

Figure 9.18 shows the surface given by fixing just θ at $\theta = \pi/4$.

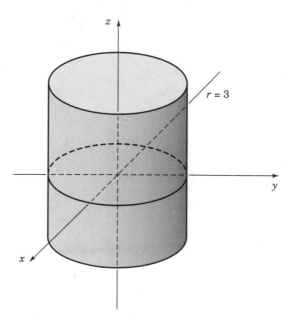

Figure 9.17 The surface described by $r = 3$.

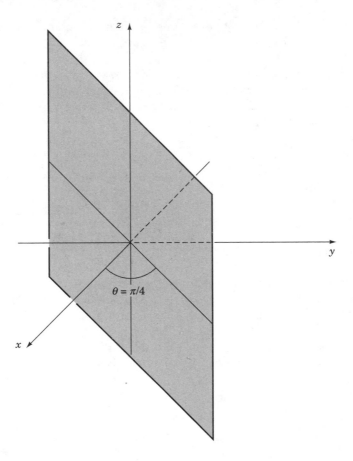

Figure 9.18 The surface described in cylindrical coordinates by $\theta = \pi/4$.

Example 2 Graph $z = r^2$.

Trick If we have an equation missing θ, it's easy to graph. Let's see how in this example.

Solution Notice that when we are in the yz-plane, and we pick a point $(0, y, z)$, y and r for that point are the same. Both are the distance from the point back to the z-axis. So for points in the yz-plane, the equation $z = r^2$ is equivalent to the equation $z = y^2$. Let's graph that parabola in the yz-plane. Now, since θ doesn't appear in the equation $z = r^2$, θ can be anything. So we just spin the parabola in the yz-plane around the z-axis. Any point on the resulting surface of revolution will have the same r-value and z-value as the

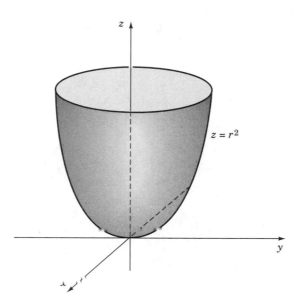

Figure 9.19 $z = r^2$ gives the paraboloid in cylindrical coordinates.

point that it came from in the yz-plane, and so it will satisfy the equation $z = r^2$. The result is a bowl, as in Figure 9.19. In fact, we have seen this bowl before. It's just the most generic paraboloid.

Another way to see that $z = r^2$ generates the generic paraboloid is by substituting $x^2 + y^2$ for r^2. Then $z^2 = x^2 + y^2$ is the equation of the most standard paraboloid.

INTEGRALS IN CYLINDRICAL COORDINATES

Triple integrals over regions which are described by their distance from a line, like cylinders and cones, are often best done by converting to cylindrical coordinates.

Since cylindrical coordinates are just polar coordinates with a z added on, the volume of a tiny box is its area in the $r\theta$-plane times its height. In the limit, this implies $dV = r\, dr\, d\theta\, dz$. So we get:

The formula for integrating a function f over a solid region V in cylindrical coordinates is

$$\iiint\limits_V f(r, \theta, z)\, r\, dr\, d\theta\, dz$$

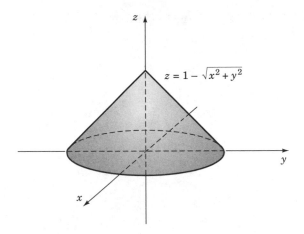

Figure 9.20 The region under $z = 1 - \sqrt{x^2 + y^2}$.

Example 3   Calculate the integral of $f(x, y, z) = \sqrt{x^2 + y^2}$ over the solid S which lies above the plane $z = 0$ and below the surface $z = 1 - \sqrt{x^2 + y^2}$.

Solution   The surface $z = 1 - \sqrt{x^2 + y^2}$ describes an upside-down cone. A picture of the solid S appears in Figure 9.20.

Engineers, physicists, and children who have just dropped their ice cream cones have to do calculations like this all the time. So how do we do it? In Cartesian coordinates it's a real mess, with square roots all over the place. But in cylindrical coordinates it's a cinch. The nasty term $f(x, y, z) = \sqrt{x^2 + y^2}$ converts to the nice term $f(r, \theta, z) = r$. The lower- and upper-bounding surfaces for the solid in polar coordinates are $z = 0$ and $z = 1 - r$, since $1 - \sqrt{x^2 + y^2} = 1 - r$. The solid is then given by the inequalities:

$$0 \le z \le 1 - r$$
$$0 \le r \le 1$$
$$0 \le \theta \le 2\pi$$

So the triple integral for this problem is set up as

$$\int_0^{2\pi} \int_0^1 \int_0^{1-r} rr\, dz\, dr\, d\theta$$

We evaluate this iteratively, starting with the inside integral and working outward.

$$\int_0^{2\pi} \int_0^1 \int_0^{1-r} r^2 \, dz \, dr \, d\theta = \int_0^{2\pi} \int_0^1 \left[r^2 z \right]_{z=0}^{z=1-r} dr \, d\theta$$

$$= \int_0^{2\pi} \int_0^1 r^2(1-r) \, dr \, d\theta$$

$$= \int_0^{2\pi} \int_0^1 (r^2 - r^3) \, dr \, d\theta$$

$$= \int_0^{2\pi} \left[\frac{r^3}{3} - \frac{r^4}{4} \right]_0^1 d\theta$$

$$= \int_0^{2\pi} (\tfrac{1}{3} - \tfrac{1}{4}) \, d\theta$$

$$= \tfrac{1}{12} \left[\theta \right]_0^{2\pi}$$

$$= \frac{\pi}{6}$$

SPHERICAL COORDINATES

Cylindrical coordinates were fun, but they were just an extended version of polar coordinates. You're probably saying, "How about a really new coordinate system? One that I can use to impress my new love interest?" You're in luck because we have just the thing.

Given a point in 3-space, connect it to the origin with a line segment, as in Figure 9.21. The length of this line segment is called ρ (rho, pronounced "row"). Greek letters always make for more impressive coordinate systems. The angle that the line segment makes with the positive z-axis is called ϕ (phi, pronounced "fee"). If the value of ϕ is 0, the point is straight up the z-axis. If $\phi = \pi$, then the point is down on the negative z-axis. If $\phi = \pi/2$, then the point is at a right angle to the z-axis, on the xy-plane. Now project the line segment into the xy-plane and define the angle of that projected line segment with the positive x-axis to be θ. This θ is the same as our old pal from cylindrical coordinates.

Usually when we specify a point we take ρ to be nonnegative, ϕ to be between 0 (north pole) and π (south pole), and θ to be between 0 and 2π. These three coordinates (ρ, ϕ, θ) completely determine where a point is in space. They are its *spherical coordinates*. The angles θ and ϕ specify a direction from the origin, or a point on a sphere, and the radius ρ specifies the distance of a point from the origin.

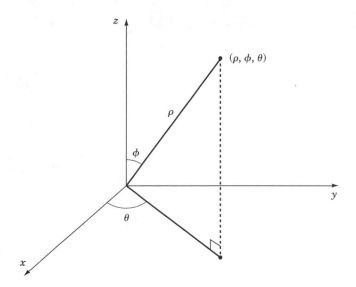

Figure 9.21 Spherical coordinates.

Warning Some people will just use r for ρ. They may even switch the meaning of ϕ and θ or write (ρ, θ, ϕ) instead of (ρ, ϕ, θ). But our notation is the most common.

These coordinates work really well for describing certain shapes, namely, spheres. They also work well for a few other shapes like ellipsoids, but spheres are number one. Calculations on spheres come up a lot, which is why spherical coordinates are so handy. You'll see them when calculating the electric field around a particle, the gravitational field around a planet, and the mass of a water balloon.

There are simple rules for converting from spherical to rectangular coordinates:

$$
\begin{aligned}
x &= \rho \sin \phi \cos \theta \\
y &= \rho \sin \phi \sin \theta \\
z &= \rho \cos \phi
\end{aligned}
$$

Each of these equations can be read right off Figure 9.22.

Notice that the angle ϕ is not only the angle we defined it to be, but it is also the angle at the top of the vertical right triangle by elementary geometry. Once we see that, we know $z = \rho \cos \phi$ from the vertical right triangle. We also see $r = \rho \sin \phi$. Then we have

$$
x = r \cos \theta = (\rho \sin \phi) \cos \theta
$$

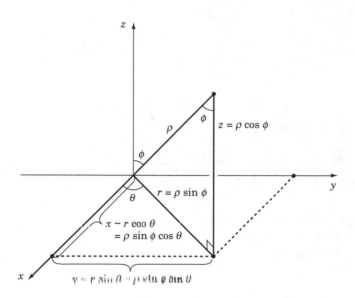

Figure 9.22 Converting spherical coordinates to rectangular coordinates.

and

$$y = r \sin \theta = (\rho \sin \phi) \sin \theta$$

Example 4   What are the rectangular coordinates for the point whose spherical coordinates are $(10, \pi/3, \pi/2)$?

Solution   Well, we just plug into the formulas.

$$x = \rho \sin \phi \cos \theta = (10)\left(\sin \frac{\pi}{3}\right)\left(\cos \frac{\pi}{2}\right) = (10)\left(\frac{\sqrt{3}}{2}\right)(0) = 0$$

$$y = \rho \sin \phi \sin \theta = (10)\left(\frac{\sqrt{3}}{2}\right)(1) = 5\sqrt{3}$$

$$z = \rho \cos \phi = (10)\left(\cos \frac{\pi}{3}\right) = (10)\left(\frac{1}{2}\right) = 5$$

So the rectangular, or Cartesian, coordinates of the point are $(0, 5\sqrt{3}, 5)$.

Converting from rectangular to spherical is slightly trickier. Suppose we know (x, y, z) and want to work out (ρ, ϕ, θ). We get ρ from the equation

$$\rho^2 = x^2 + y^2 + z^2$$

which is the distance formula from the origin to (x, y, z). We can then work out ϕ from $z = \rho \cos \phi$, since we already know z and ρ. Finally we can get the value of θ from $x = \rho \sin \phi \cos \theta$, since the other quantities are all known. With this last step we need to take care to put θ in the correct quadrant.

Example 5   What are the spherical coordinates for the point whose rectangular coordinates are $(1, 1, \sqrt{2})$?

Solution   We calculate that

$$\rho^2 = x^2 + y^2 + z^2 = 1^2 + 1^2 + 2 = 4$$

So $\rho = 2$. Next we get ϕ from $z = \sqrt{2}$, which gives

$$z = \rho \cos \phi$$
$$\sqrt{2} = 2 \cos \phi$$
$$\cos \phi = \frac{\sqrt{2}}{2} = \frac{1}{\sqrt{2}}$$
$$\phi = \frac{\pi}{4}$$

Finally, we calculate θ from $x = \rho \sin \phi \cos \theta$:

$$1 = 2 \left(\frac{\sqrt{2}}{2} \right) \cos \theta$$
$$\cos \theta = \frac{1}{\sqrt{2}}$$
$$\theta = \frac{\pi}{4}$$

[We have to be careful to pick θ in the correct quadrant, since $\cos(7\pi/4)$ also equals $1/\sqrt{2}$. But the xy-coordinates show that θ is in the first quadrant.]
    So the spherical coordinates of the point are $(\rho, \phi, \theta) = (2, \pi/4, \pi/4)$.

INTEGRALS IN SPHERICAL COORDINATES

When computing a triple integral in spherical coordinates, we need to know what $dV = dx \, dy \, dz$ should become in spherical coordinates. We determine this by varying each of our spherical coordinates just a bit, and seeing what the formula is for the volume of the resulting solid. Letting ρ vary by $\Delta\rho$, ϕ vary by $\Delta\phi$, and θ vary by $\Delta\theta$, we get the solid that appears in Figure 9.23.

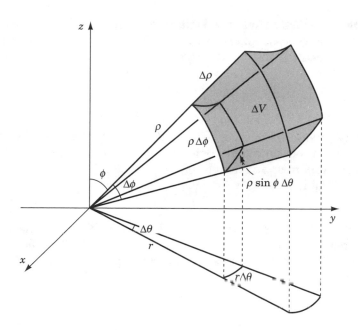

Figure 9.23 Varying each coordinate in spherical coordinates.

This object is almost a box, with a width of $r\,\Delta\theta = \rho\sin\phi\,\Delta\theta$, a radial length of $\Delta\rho$, and a depth of $\rho\,\Delta\phi$. So its volume is given approximately by

$$\Delta V \approx (\rho\sin\phi\,\Delta\theta)(\Delta\rho)(\rho\,\Delta\phi)$$

$$\approx \rho^2\sin\phi\,\Delta\rho\,\Delta\phi\,\Delta\theta$$

In the limit, as we cut up a solid into tinier and tinier pieces, this approximation becomes exact. So

$$\iiint_V f(x, y, z)\,dV = \iiint_V f(\rho, \phi, \theta)\,\rho^2\sin\phi\,d\rho\,d\phi\,d\theta$$

Anytime we have an integral in spherical coordinates, we automatically stick in the $\rho^2\sin\phi\,d\rho\,d\phi\,d\theta$ for dV.

Example 6 (Eliminating Air Pollution)   The Xtilploth aliens are siphoning off the Earth's atmosphere to take to their own oxygen-depleted world. They need to calculate the atmosphere's total mass to figure out how long it will take to them to tow it home. For rough estimation, they assume the Earth is a sphere of radius 6,400,000 meters, and that the density of air h meters above the ground is $\dfrac{10^6}{6,400,000 + h}$ kilograms per cubic meter. (*Note:* Don't use this in your meteorology class, it's not even close. But hey, they're aliens, who can fathom how they come up with these functions?) If they take all of the atmosphere up to a height of 20 kilometers, what is the total mass of the air they tug home?

Solution   We calculate the total mass of the atmosphere by integrating its density. (More on that in the next section.) This is best done in spherical coordinates. We'll measure everything in meters and kilograms. The region we want to integrate over is described in spherical coordinates as having

$$6,400,000 \leq \rho \leq 6,420,000$$

(6400 to 6420 kilometers from the center of Earth)

$$0 \leq \phi \leq \pi$$

and

$$0 \leq \theta \leq 2\pi$$

Since $h = \rho - 6,400,000$, the function we integrate is

$$f(\rho, \phi, \theta) = \frac{10^6}{6,400,000 + (\rho - 6,400,000)}$$

$$= \frac{10^6}{\rho}$$

which gives the atmosphere's density at a point (ρ, ϕ, θ). So the triple integral we want is

$$\int_0^{2\pi} \int_0^{\pi} \int_{6,400,000}^{6,420,000} \frac{10^6}{\rho} \, \rho^2 \sin \phi \, d\rho \, d\phi \, d\theta$$

Doing this integral from the inside out, we find:

$$\int_0^{2\pi} \int_0^{\pi} \int_{6,400,000}^{6,420,000} \frac{10^6}{\rho} \rho^2 \sin\phi \, d\rho \, d\phi \, d\theta = 10^6 \int_0^{2\pi} \int_0^{\pi} \int_{6,400,000}^{6,420,000} \rho \sin\phi \, d\rho \, d\phi \, d\theta$$

$$= 10^6 \int_0^{2\pi} \int_0^{\pi} \left[\frac{\rho^2}{2} \right]_{6,400,000}^{6,420,000} \sin\phi \, d\phi \, d\theta$$

$$\approx 10^6 \int_0^{2\pi} \int_0^{\pi} 1.282 \times 10^{11} \sin\phi \, d\phi \, d\theta$$

$$= 1.282 \times 10^{17} \int_0^{2\pi} \int_0^{\pi} \sin\phi \, d\phi \, d\theta$$

$$= 1.282 \times 10^{17} \int_0^{2\pi} \left[-\cos\phi \right]_0^{\pi} \, d\theta$$

$$= 1.282 \times 10^{17} \int_0^{2\pi} \left[-(-1) - (-1) \right] d\theta$$

$$= 1.282 \times 10^{17} \left[2\theta \right]_0^{2\pi}$$

$$= 1.282 \times 10^{17} (4\pi)$$

$$\approx 1.611 \times 10^{18} \text{ kg}$$

This is too much mass for them to drag home in time. No problem, though. They sent it air freight.

9.7 Mass, center of mass, and moments

In math and physics, we often pretend that a messily shaped object is just a single point. If we sling a cow across a pasture using a catapult, and we want to make sure it will land in the pond and not on the local supermarket, then it doesn't make too much difference if we assume the cow is a point, with all of its mass centered at that point. What is important is to know the total mass of the cow and the so-called *center of mass* of the cow.

MASS

First, how to find the mass of a cow V (stands for "veal")? Suppose that the cow's density is not uniform. Certain parts of the cow are denser than other parts. And suppose that the density of the cow at a point (x, y, z) is given by $\sigma(x, y, z)$. Then if we cut the cow up into lots of little boxes (not literally, just figuratively; we're not Oscar Meyer), the mass of a box is approximately the density at some point in the box times the volume of the box. Adding up the masses of all these boxes and taking a limit as the number of boxes goes up and the size of the boxes goes down, we get an integral:

$$\text{Mass} = \iiint\limits_{V} \sigma(x, y, z)\, dx\, dy\, dz$$

Notice that when the density $\sigma(x, y, z) = 1$, then the mass is just equal to the volume.

CENTER OF MASS

The center of mass is also called the *centroid,* and to understand it, let's return to the childhood days you spent at the playground. Remember how your favorite thing in the world was to get on one end of a seesaw, with Bobby Rensputter at the other end, and to slowly bounce up and down, while singing "See-Saw, marjory daw, the pumpkin went to the market" or some such ditty, until Bobby was lulled into a false sense of security. And then when he was right at the highest point, you rolled off your end of the seesaw and watched in delight as that look of despair crossed his face and he plummeted to the teeth-jarring finale. Yes, childhood was a constant parade of delights.

And those experiences on the seesaw taught you something about the mechanics of balancing and fulcrums. They also taught Bobby Rensputter about the darker side of human nature. But let's stick to what you learned.

THEORY OF SEESAWS

Suppose that one of the players from the offensive front line of the Green Bay Packers sits down on a reinforced seesaw at x_1, with a member of the New York City Ballet at the other end at x_2. Needless to say, it had better be the case that the balancing point $\bar{x}$ is nearer to the Packer than the ballerina. (See Figure 9.24.)

Figure 9.24 A seesaw.

In particular, it must be the case that if $\bar{x}$ is the balancing point, then

$$m_1(\bar{x} - x_1) = m_2(x_2 - \bar{x})$$

On each side, the mass times the distance of the mass from the balance point are equal.

Solving that equation for $\bar{x}$, we get

$$m_1\bar{x} - m_1x_1 = m_2x_2 - m_2\bar{x}$$
$$m_1\bar{x} + m_2\bar{x} = m_1x_1 + m_2x_2$$
$$(m_1 + m_2)\bar{x} = m_1x_1 + m_2x_2$$

$$\boxed{\bar{x} = \frac{m_1x_1 + m_2x_2}{m_1 + m_2}}$$

The coordinate of the balance point is obtained by taking a weighted sum of the masses divided by the total mass. The masses are each weighted by their respective coordinate.

For a whole bunch of masses, say n, the formula would be

$$\boxed{\bar{x} = \frac{\sum_{i=1}^{n} m_i x_i}{\sum_{i=1}^{n} m_i} = \frac{m_1x_1 + m_2x_2 + \cdots + m_nx_n}{m_1 + m_2 + \cdots + m_n}}$$

Now this is just on a line, but the same rule applies for the y- and z-coordinates when our particles are distributed in space.

$$\boxed{\bar{y} = \frac{\sum_{i=1}^{n} m_i y_i}{\sum_{i=1}^{n} m_i}}$$

$$\boxed{\bar{z} = \frac{\sum_{i=1}^{n} m_i z_i}{\sum_{i=1}^{n} m_i}}$$

The point $(\bar{x}, \bar{y}, \bar{z})$ is called the centroid, or center of mass, of the n particles.

Our next goal is to find the center of mass of any old solid body V. The idea for finding it is to cut up the body into little bits, and treat them as a whole lot of points whose center of mass we want to find. To compute the mass of a little box in the region, we need to know the distribution of mass in the region, which is given by a density function $\sigma(x, y, z)$. We multiply the volume of that region ($dxdydz$ after a limit) by the density $\sigma(x, y, z)$. This leads us to three integral formulas:

$$\bar{x} = \frac{\iiint_V \sigma(x, y, z) \, x \, dx \, dy \, dz}{\iiint_V \sigma(x, y, z) \, dx \, dy \, dz}$$

$$\bar{y} = \frac{\iiint_V \sigma(x, y, z) \, y \, dx \, dy \, dz}{\iiint_V \sigma(x, y, z) \, dx \, dy \, dz}$$

$$\bar{z} = \frac{\iiint_V \sigma(x, y, z) \, z \, dx \, dy \, dz}{\iiint_V \sigma(x, y, z) \, dx \, dy \, dz}$$

Notice that in all three cases, the triple integral in the denominator is just the mass of the solid. The point $(\bar{x}, \bar{y}, \bar{z})$ is the center of mass of the body, and it's right where all the mass would balance if we hung that region from a string.

Because we multiply the density function by the first power of x, y, or z to calculate these, the integrals in the numerator are called the *first moments* of the object. This does not refer to when the object was born.

Trick We can often use symmetry to simplify center of mass calculations. If the body you are looking at looks exactly the same when you switch x and $-x$ and the density function σ is unchanged by this switch, then the x-coordinate of the center of mass must be zero. Similarly for y and z.

Example 1 Find the centroid of the solid with constant density $\sigma(x, y, z) = 1$ that is bounded by the graphs of $z = x^2 + y^2$ and $z = 4$.

Solution The solid is drawn in Figure 9.25. Both the solid and the density function are unchanged when we replace x by $-x$. Similarly when we replace y by $-y$. So we have $\bar{x} = 0$ and $\bar{y} = 0$. But the region is not unchanged when z is replaced by $-z$, so we have to calculate some integrals to find $\bar{z}$.

$$\bar{z} = \frac{\iiint_V \sigma(x, y, z) \, z \, dx \, dy \, dz}{\iiint_V \sigma(x, y, z) \, dx \, dy \, dz}$$

Let's calculate each of the numerator and denominator, starting with the denominator, which gives the total mass. We'll use cylindrical coordinates. The region is described by

$$r^2 \leq z \leq 4$$
$$0 \leq r \leq 2$$
$$0 \leq \theta \leq 2\pi$$

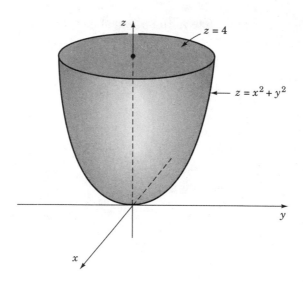

Figure 9.25 Finding the center of mass.

So the triple integral for the mass is

$$\int_0^{2\pi} \int_0^2 \int_{r^2}^4 r\, dz\, dr\, d\theta$$

We evaluate this iteratively, starting with the inside integral and working outward.

$$
\begin{aligned}
\text{Mass} &= \int_0^{2\pi} \int_0^2 \int_{r^2}^4 r\, dz\, dr\, d\theta \\
&= \int_0^{2\pi} \int_0^2 \Big[rz \Big]_{z=r^2}^{z=4} dr\, d\theta \\
&= \int_0^{2\pi} \int_0^2 [4r - r^3]\, dr\, d\theta \\
&= \int_0^{2\pi} \left[2r^2 - \frac{r^4}{4} \right]_0^2 d\theta \\
&= \int_0^{2\pi} 4\, d\theta \\
&= 8\pi
\end{aligned}
$$

The numerator is given by the integral

$$\int_0^{2\pi} \int_0^2 \int_{r^2}^4 zr\, dz\, dr\, d\theta = \int_0^{2\pi} \int_0^2 \left[\left(\frac{z^2}{2}\right)r\right]_{z=r^2}^{z=4} dr\, d\theta$$

$$= \int_0^{2\pi} \int_0^2 \left[8r - \frac{r^5}{2}\right] dr\, d\theta$$

$$= \int_0^{2\pi} \left[4r^2 - \frac{r^6}{12}\right]_0^2 d\theta$$

$$= \int_0^{2\pi} \frac{32}{3}\, d\theta$$

$$= \frac{64}{3}\pi$$

So

$$\bar{z} = \frac{(64/3)\pi}{8\pi} = \frac{8}{3}$$

and the center of mass is at $(0, 0, 8/3)$.

Example 2   Find the center of mass of a ball of radius 1 whose density is given by $\sigma(x, y, z) = 3x^2 + 2y^4 + z^6$.

Solution   No problem. Since the body we are studying is the same at (x, y, z) and $(-x, y, z)$, and the density is the same at (x, y, z) and $(-x, y, z)$, we must have that $\bar{x} = 0$. For similar reasons, $\bar{y} = 0$ and $\bar{z} = 0$. So the center of mass is at $(0, 0, 0)$.

MOMENT OF INERTIA

What is the moment of inertia? You know it well. That instant when the person you had a crush on in calculus class passed you in the hall and looked right at you, eyes like rayguns firing deep into your soul. There was a split second of opportunity, which if seized, could have led to a tandem bicycle for two down the road of life. You wanted to say, "Hi, I'm that person who stares at the back of your head all class long." But what came out was a garbled, "Egurrhhh." You watched horrified as your potential soulmate walked off with a confused backward glance. The bicycle built for two careened wildly into a ditch. That was your "moment of inertia," a term very familiar to those in the "hard" sciences.

But we want to talk about another type of moment of inertia.

Class Demonstration   Split up into pairs. Pick up your neighbor and twirl him or her around in the air. Be careful not to let go. Note that it's hard work to get the motion going at first, but once you get your partner spinning good, with feet up in the air, it's not so bad. The resistance to getting a body twirling is what the moment of inertia measures.

For a particle, the moment of inertia is given by mr^2, where m is the mass and r is the distance to the axis we are swinging the particle around. Because we multiply the mass by the second power of r to calculate this, it's also called the *second moment* of the object.

To calculate the moment of inertia for an object which fills up a solid V, we cut it up into tiny little boxes. The mass per unit volume at a point is given by a function called the mass density, so the mass of a tiny box is approximated well by multiplying its volume by the mass density. The moment of inertia of the tiny box is then its mass times the square of its distance from the axis. As usual, this results in an integral. Let's say we are rotating around the z-axis. Then the square of the distance from the axis is given by $x^2 + y^2$ and the moment of inertia of a solid V with mass density $\sigma(x, y, z)$ is

$$\iiint_V \sigma(x, y, z)(x^2 + y^2)\, dx\, dy\, dz$$

or in cylindrical coordinates,

$$\iiint_V \sigma(r, \theta, z)(r^2)\, r\, dr\, d\theta\, dz$$

Besides calculating the amount of energy needed to start an engine turning, the moment of inertia is also important in the study of animal tipping.

Example 3 (Chicken Spinning)   Having built her bronze chicken sculpture, Wynodia propped it 2 meters off the ground on a thin steel rod, so that it now occupies the region under $z = 11 - x^2 - y^2$ and above $z = 2$. Wynodia decides that she wants to spin her chicken like a top on its single leg, and to change the sculpture's name to "chicken spin." (The workers nickname it "spitted chicken barbecue.") To ensure it will spin easily, calculate the moment of inertia about the z-axis. The density of bronze is 8900 kilograms per cubic meter, so $\sigma(x, y, z) = 8900$.

Solution   The moment of inertia around the z-axis of the chicken is given by the cylindrical coordinates integral:

$$\int_0^3 \int_0^{2\pi} \int_2^{11-r^2} (8900)r^2 r \, dz \, d\theta \, dr$$

Notice that we have set it up to integrate with respect to z first, then θ, then r. We can choose any order to do multiple integrals, and this one is best suited for this problem.

Then the integral equals

$$\int_0^3 \int_0^{2\pi} \int_2^{11-r^2} (8900)r^3 \, dz \, d\theta \, dr = \int_0^3 \int_0^{2\pi} (8900)[r^3 z]_{z=2}^{z=11-r^2} \, d\theta \, dr$$

$$= \int_0^3 \int_0^{2\pi} (8900)(9r^3 - r^5) \, d\theta \, dr$$

$$= \int_0^3 (2\pi)(8900)(9r^3 - r^5) \, dr$$

$$= (2\pi)(8900)\left[\frac{9r^4}{4} - \frac{r^6}{6}\right]_0^3$$

$$= (2\pi)(8900)\left[\frac{9(3^4)}{4} - \frac{3^6}{6}\right]$$

$$\approx 3.4 \times 10^6$$

That's a lot of inertia. She'd better get some well-oiled ball bearings if she wants to get that chicken spinning.

9.8 Change of coordinates

We have already seen how to calculate multiple integrals in polar, cylindrical, and spherical coordinates. But there are an unlimited number of different coordinate systems possible. You can even make your own up. But how do we decide what replaces $dV = dx \, dy \, dz$ in some random coordinate system? If we want to do integration in these new coordinates, we are going to need to know that. Luckily, there is a simple formula which tells us.

Suppose that a point in space is described by coordinates u, v, and w, rather than the usual rectangular coordinates x, y, and z. The new coordinates are related to x, y, z by equations $x = x(u, v, w)$, $y = y(u, v, w)$, $z = z(u, v, w)$. Then the formula for the triple integral of a function $f(x, y, z)$, $\int \int \int_V f(x, y, z) \, dV$ will be replaced by $\int \int \int_V f(u, v, w) \, dV$. What we need to do is to express dV

in our new coordinates, by finding the volume of a small box-shaped region in the u-, v-, w-coordinate system. Set the vector $\mathbf{r}$ to be $x\mathbf{i} + y\mathbf{j} + z\mathbf{k}$. Then $\mathbf{r} = x(u, v, w)\mathbf{i} + y(u, v, w)\mathbf{j} + z(u, v, w)\mathbf{k}$. The volume of a small box in u-, v-, w-coordinates is given approximately by the scalar triple product of the vectors

$$\frac{\partial \mathbf{r}}{\partial u}, \frac{\partial \mathbf{r}}{\partial v}, \text{ and } \frac{\partial \mathbf{r}}{\partial w}$$

The formula for dV is given by the scalar triple product of these, or

$$dV = \begin{vmatrix} \dfrac{\partial x}{\partial u} & \dfrac{\partial y}{\partial u} & \dfrac{\partial z}{\partial u} \\[2mm] \dfrac{\partial x}{\partial v} & \dfrac{\partial y}{\partial v} & \dfrac{\partial z}{\partial v} \\[2mm] \dfrac{\partial x}{\partial w} & \dfrac{\partial y}{\partial w} & \dfrac{\partial z}{\partial w} \end{vmatrix} du\, dv\, dw = \left| \frac{\partial(x, y, z)}{\partial(u, v, w)} \right| du\, dv\, dw$$

The last expression is just a shorthand way of writing this. This determinant is called the *Jacobian*; it measures the volume stretching factor of u, v, w compared to x, y, z.

The triple integral in u-, v-, w-coordinates is then written as

$$\iiint\limits_{V} f(u, v, w) \left| \frac{\partial(x, y, z)}{\partial(u, v, w)} \right| du\, dv\, dw$$

V will be described using the u-, v-, w-coordinates which correspond to the region we are integrating over. Note that this is not the same as the description of that region in x-, y-, z-coordinates.

Example 1 (Feet and Yard Coordinates)   The cubicles smelled. Too many people had been eating too many late-night snacks and stuffing the garbage behind a neighbor's desk. So everyone was required to measure the size of his or her own cubicles in cubic feet, and buy enough air freshener to get rid of the odor. Air freshener costs $5 per cubic foot (the strong stuff), so most people described their cubicles in units of feet and integrated the function $f(x, y, z) = 5$ over the region describing their cubicle to calculate the cost of the air freshener they needed. As usual however, Amy did things differently. Amy always insisted on measuring in yards rather than feet. Her cubicle S, in the usual x-, y-, z-foot coordinates, was the region $0 \le x \le 6, 0 \le y \le 6, 0 \le z \le 6$ (cubical of course). But using yard coordinates (u, v, w), with $x = 3u$,

$y = 3v$, and $w = 3z$, Amy measured the cubicle as $0 \le u \le 2$, $0 \le v \le 2$, $0 \le w \le 2$. Set up the triple integral and measure the cost of freshening Amy's cubicle using u-, v-, w-coordinates.

Solution We calculate the Jacobian for changing yards (u, v, w) to feet (x, y, z). Since this involves multiplying each coordinate by 3, we have

$$\mathbf{r} = x(u, v, w)\mathbf{i} + y(u, v, w)\mathbf{j} + z(u, v, w)\mathbf{k} = 3u\mathbf{i} + 3v\mathbf{j} + 3w\mathbf{k}$$

We calculate

$$\frac{\partial \mathbf{r}}{\partial u} = 3\mathbf{i}$$

$$\frac{\partial \mathbf{r}}{\partial v} = 3\mathbf{j}$$

$$\frac{\partial \mathbf{r}}{\partial w} = 3\mathbf{k}$$

The Jacobian is the scalar triple product of these vectors, or

$$\begin{vmatrix} 3 & 0 & 0 \\ 0 & 3 & 0 \\ 0 & 0 & 3 \end{vmatrix} = 27$$

The region of integration S in u-, v-, w-coordinates is $0 \le u \le 2$, $0 \le v \le 2$, $0 \le w \le 2$. So the triple integral is

$$\iiint\limits_{S} f(u, v, w) \left| \frac{\partial(x, y, z)}{\partial(u, v, w)} \right| du \, dv \, dw = \int_0^2 \int_0^2 \int_0^2 5 \cdot 27 \, du \, dv \, dw$$

which evaluates to \$1080. Yikes. And it wouldn't help her if she computed in x-, y-, z-coordinates. She would get the same answer. Hope the salary's good.

Example 2 (Rectangular to Spherical Coordinates) Calculate the Jacobian for changing rectangular coordinates (x, y, z) to spherical coordinates (ρ, ϕ, θ), and use it to calculate the volume of a ball of radius R in spherical coordinates.

Solution We have $x = \rho \sin \phi \cos \theta$, $y = \rho \sin \phi \sin \theta$, $z = \rho \cos \phi$. So

$$\mathbf{r} = \rho \sin \phi \cos \theta \mathbf{i} + \rho \sin \phi \sin \theta \mathbf{j} + \rho \cos \phi \mathbf{k}$$

and we calculate

$$\frac{\partial \mathbf{r}}{\partial \rho} = \sin\phi\cos\theta\mathbf{i} + \sin\phi\sin\theta\mathbf{j} + \cos\phi\mathbf{k}$$

$$\frac{\partial \mathbf{r}}{\partial \phi} = \rho\cos\phi\cos\theta\mathbf{i} + \rho\cos\phi\sin\theta\mathbf{j} - \rho\sin\phi\mathbf{k}$$

$$\frac{\partial \mathbf{r}}{\partial \theta} = -\rho\sin\phi\sin\theta\mathbf{i} + \rho\sin\phi\cos\theta\mathbf{j}$$

The Jacobian is the scalar triple product of these vectors, or

$$\begin{vmatrix} \sin\phi\cos\theta & \sin\phi\sin\theta & \cos\phi \\ \rho\cos\phi\cos\theta & \rho\cos\phi\sin\theta & -\rho\sin\phi \\ -\rho\sin\phi\sin\theta & \rho\sin\phi\cos\theta & 0 \end{vmatrix} = \rho^2\sin\phi$$

Hey, we saw this expression when we discussed triple integrals in spherical coordinates. Isn't it reassuring when math is consistent?

The volume of the ball of radius R is obtained by integrating this Jacobian over the region where $0 \le \rho \le R, 0 \le \phi \le \pi, 0 \le \theta \le 2\pi$.

This gives

$$\int_0^{2\pi} \int_0^{\pi} \int_0^R \rho^2 \sin\phi \, d\rho \, d\phi \, d\theta = \int_0^{2\pi} \int_0^{\pi} \frac{R^3}{3} \sin\phi \, d\phi \, d\theta$$

$$= \int_0^{2\pi} \frac{R^3}{3}[-\cos\phi]_0^{\pi} \, d\phi \, d\theta$$

$$= \int_0^{2\pi} \frac{2R^3}{3} \, d\theta$$

$$= \frac{4}{3}\pi R^3$$

That certainly is the famous formula for the volume of a ball of radius R.

Hey, whatever staffs your help desk.

Vector Fields and the Green-Stokes Gang

 Vector fields

A vector is like a single arrow, all by itself. But if a farmer were to grow arrows, using some genetically modified wheat, then there would be an arrow at every point in the farmer's land, and this is called a *vector field*. In fact, these growing arrows are very heavy, so they don't stand up straight—they lie on the ground. A vector field on the plane gives a two-dimensional vector at each point in the plane.

An example of a simple two-dimensional vector field that we experience every day is the wind. Suppose we're walking around our farm on a windy day. At each point on the ground the wind is blowing with a certain speed and horizontal direction. Since direction and speed are captured by a vector, we can understand which way the wind blows by giving a two-dimensional vector at each point on the ground. We can picture such a vector field by drawing the vectors at each of the points in the plane, as in Figure 10.1.

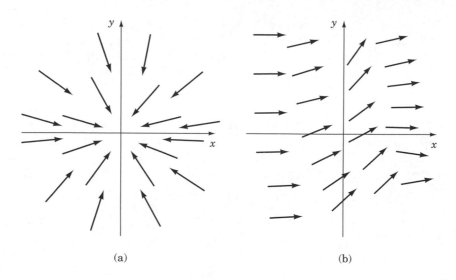

(a)   (b)

Figure 10.1 Examples of vector fields in the plane.

In equations we write

$$\mathbf{W}(x, y) = P(x, y)\mathbf{i} + Q(, y)\mathbf{j}$$

This vector field assigns to each point in the xy-plane a two-dimensional vector. Notice we are using the $\mathbf{i}, \mathbf{j}$ notation for vectors. That's because this is the notation favored by physicists, and physics is where vector fields rule.

If you're at the point where $x = 3$ and $y = 1$ (somewhere in Oklahoma) and the wind vector field is given by the formula

$$\mathbf{W}(x, y) = (x^2 - 5y)\mathbf{i} + (6y - x)\mathbf{j}$$

then the wind that you feel on your feet is described by the vector $4\mathbf{i} + 3\mathbf{j}$. The magnitude of this vector, calculated as $\sqrt{4^2 + 3^2} = 5$, gives the speed of the wind. The direction of the wind is given by the direction of the vector $4\mathbf{i} + 3\mathbf{j}$. If the wind knocked you down, that's the direction your body would point when it landed.

We can also have a vector field in space. This gives a three-dimensional vector $\mathbf{F}(x, y, x)$ at every point (x, y, z) in 3-space. If you are flying around in an airplane, then the wind vector field is described by a function with one coordinate to measure each of the x-, y-, and z-directions. It looks like

$$\mathbf{F}(x, y, z) = P(x, y, z)\mathbf{i} + Q(x, y, z)\mathbf{j} + R(x, y, z)\mathbf{k}$$

Besides wind, vector fields are used to describe forces, like electric force, gravitational force, and magnetic force. Forces, like bullies, push you around if you let them. At each point in space, the strength and direction in which the force is pushing is given by a vector, so force is a vector field.

Example (Gravitational Vector Field) If the Earth is centered at the origin $(0, 0, 0)$, its gravitational force sucks everything around it down toward its center. This is what stops Australians from falling off the bottom and interfering with communication satellites. The gravitational force field is given by

$$\mathbf{F}(x, y, z) = \left(\frac{-cx}{(x^2 + y^2 + z^2)^{3/2}} \right) \mathbf{i} + \left(\frac{-cy}{(x^2 + y^2 + z^2)^{3/2}} \right) \mathbf{j} + \left(\frac{-cz}{(x^2 + y^2 + z^2)^{3/2}} \right) \mathbf{k}$$

where c is a constant depending on which units we use. This looks messy, but as anyone who drops a tomato or an egg can testify, gravity is like that. It's not too bad when you consider it tells how objects are affected by the Earth's gravity no matter where they are. If a unit of mass is located at the point $(0, 10,000, 0)$, then the gravitational vector field exerts a force on it given by the vector

$$F(0, 10,000, 0) = \frac{-c(0)}{((0)^2 + (10,000)^2 + (0)^2)^{3/2}} \mathbf{i} + \frac{-c(10,000)}{((0)^2 + (10,000)^2 + (0)^2)^{3/2}} \mathbf{j}$$

$$+ \frac{-c(0)}{((0)^2 + (10,000)^2 + (0)^2)^{3/2}} \mathbf{k}$$

$$= \frac{-c(10,000)}{(10,000)^3} \mathbf{j} = \frac{-c}{(10,000)^2} \mathbf{j}$$

This illustrates the inverse square property of gravity—the force drops off like the distance squared.

GRADIENT FIELDS

One way to create a vector field is to take the gradient of a function. At each point, the gradient vector of the function gives a vector which is part of the resulting vector field. Taking the gradient of the function $f(x, y) = xy - x^2$ gives the vector field

$$\nabla f = \frac{\partial f}{\partial x} \mathbf{i} + \frac{\partial f}{\partial y} \mathbf{j} = (y - 2x)\mathbf{i} + (x)\mathbf{j}$$

Vector fields obtained in this way are called *conservative*, and not because they wear gray suits or blouses buttoned up to a lace collar. As we'll see soon, moving all the way around a closed curve against a conservative force field conserves energy, and the net result is no work. Everyone is intrigued by the idea of no work, so you can see why conservative vector fields are popular.

The function whose gradient we take to get a conservative vector field is called a *potential function*. So when your basketball coach says, "You have a lot of potential," it may just mean that you're a good candidate for having your gradient taken. Potential functions are a big deal in physics, where they spend a lot of time with electric potentials and such things. Gravity is also a conservative field. It's the gradient of a function whose negative is called the *gravitational potential*, which for our example is

$$f(x, y, z) = -\frac{c}{\sqrt{x^2 + y^2 + z^2}}$$

The gradient of this is the gravitational force field we saw above.

Hey, whatever churns your butter.

Getting acquainted with div and curl

Picture yourself standing on top of a tall building on a windy day. Why would you be doing this? Well, maybe you've just been dumped by the love of your life, all the money that you invested in the stock of buggyparts.com just went down the tubes, and the dentist left a message that all the caps she put in your mouth last week turn out to contain a slow-acting toxic substance. So you've come to the roof to get a different perspective on life. And now that you're here, you're overcome with the wonder of wind. Boy it's windy up here. As you move to the side a little, the wind hits you from a different angle, and with varying strength. What's happening? Suddenly you find a reason to go on in life—trying to understand the way the wind velocity vector field **W** changes as you move around.

THE DIVOMETER

The *divergence* of the wind velocity, div **W**, measures how fast air is accumulating toward or moving away from us. If we let out a big sneeze, then the air is moving away from us, and the divergence of **W** (where our mouth is) is positive. If we suck in a great gasp of air, then the air is moving toward us, and the divergence is negative.

To measure divergence, we can use a biological testing organism we call a *divometer*. The divometer consists of a single tiny amoeba, which initially is happily chugging along doing what amoebas do. To measure divergence of a vector field at a given point, we place the amoeba at the selected point and observe it carefully through a microscope. The vectors in the vector field represent the way stuff is being pushed around at different points.

If the amoeba is being expanded and stretched, then the divergence is positive. If we see little bits of amoeba arms and amoeba legs flying away in all directions, that's a sign of positive divergence. On the other tentacle, if the amoeba is being compressed, then the divergence is negative. If we see a shrinking amoeba, with little amoeba eyes bulging out, then we know the divergence is negative. If the amoeba is getting carried along, but without being stretched or squished, we know the divergence is 0.

[*Editor's Note:* Yes, you are right, amoebas don't have arms, legs, or eyes. Apologies to the Friends of One-Celled Organisms. Please don't send letters.]

Let's take a look at the official formula for divergence of a vector field **F**.

$$\text{div } \mathbf{F} = \frac{\partial P}{\partial x} + \frac{\partial Q}{\partial y} + \frac{\partial R}{\partial z}$$

That's right, it's just the sum of the partial derivative with respect to x of the x-component of the vector field plus the partial derivative with respect to y of the y-component of the vector field plus the partial derivative with respect to z of the z-component of the vector field. It's called the *divergence* since if the vector field represented the velocity vectors of a flowing fluid, say hot fudge for instance, the divergence gives the net rate at which the hot fudge is diverging away from or toward a given point. Notice that at any point in space, div **F** is a number, so div **F** is an ordinary old function on 3-space.

Example 1   What's div **F** if $\mathbf{F} = 3x^3 z\mathbf{i} + 4xyz\mathbf{j} + yz^2\mathbf{k}$?

Solution

$$\text{div } \mathbf{F} = \frac{\partial(3x^3 z)}{\partial x} + \frac{\partial(4xyz)}{\partial y} + \frac{\partial(yz^2)}{\partial z} = 9x^2 z + 4xz + 2yz$$

Let's look at the value of div **F** at a point in 3-space, like $(1, 2, -3)$. It's $-27 - 12 - 12 = -51$. If you put an amoeba at $(1, 2, -3)$, it would be compressed inwards on itself, squashed like a, well, like an amoeba.

Common Mistake   When asked to compute div **F**, students often throw various **i**'s and **j**'s into the answer. Don't forget that div **F** is just a number at each point. Don't try to turn it into a vector field.

Example 2   Determine whether the divergence is positive, negative, or zero for the point marked by the amoeba in each of the vector fields shown in Figure 10.2.

Solution   In Figure 10.2a, we can see that if we dropped an amoeba onto the plane at the marked point, the amoeba would be stretched out. So the divergence is positive. In Figure 10.2b, although the amoeba would be pushed in the direction of the vector field, it would not be stretched or compressed, so the divergence is 0. In Figure 10.2c, the amoeba would get squeezed in on itself, so the divergence is negative.

There is a trick we can use to remember the formula for the divergence. We form a slightly warped vectorlike object called "del."

$$\nabla = \frac{\partial}{\partial x}\mathbf{i} + \frac{\partial}{\partial y}\mathbf{j} + \frac{\partial}{\partial z}\mathbf{k}$$

Even though its components are not numbers, we treat it like a vector anyway, showing it the respect it may not deserve. We call ∇ an *operator* because it makes no sense by itself. Like a surgeon, it has to have something to operate on. A surgeon with no one to operate on isn't really a surgeon at all. No patients means no income, no Porsche, no status, no proud mother, no house on the beach. Using ∇, the formula for the divergence becomes

---

Formula for the Divergence of F

$$\operatorname{div} \mathbf{F} = \nabla \cdot \mathbf{F}$$

If we write out this dot product, we see that

$$\nabla \cdot \mathbf{F} = \left(\frac{\partial}{\partial x}\mathbf{i} + \frac{\partial}{\partial y}\mathbf{j} + \frac{\partial}{\partial z}\mathbf{k}\right) \cdot (P\mathbf{i} + Q\mathbf{j} + R\mathbf{k}) = \frac{\partial P}{\partial x} + \frac{\partial Q}{\partial y} + \frac{\partial R}{\partial z}$$

MEASURING CURL WITH THE CURLOMETER

The *curl* of the wind vector field $\mathbf{W}$, curl $\mathbf{W}$, measures its spinning effect. It tells you how fast the wind is trying to spin you around, and about what axis. Curl $\mathbf{W}$ is a vector field that lines up with the axis along which the wind is trying to twirl you, and whose magnitude indicates the strength of the twirling effect.

We can measure curl by using a variation on the pinwheel (not the pinwheel that kids play with). We call our measuring device a *curlometer*. On

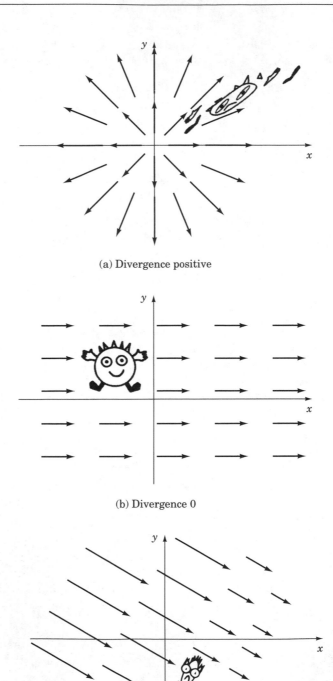

(a) Divergence positive

(b) Divergence 0

(c) Divergence negative

Figure 10.2 What's the divergence at these points?

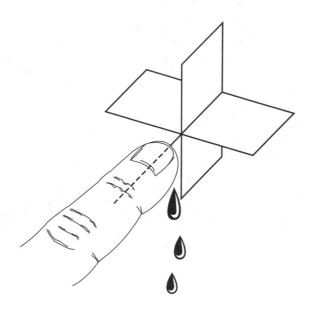

Figure 10.3 The curlometer—a device that measures curl.

a curlometer the blades catching the wind are completely flat, like the blades of a paddle wheel or a mill wheel. There's a (rather painful) way to make one, which we don't recommend, but it'll give you the idea and you won't forget it for a long time. Put the curlometer paddles on a sharp pin and stick the pin into the tip of your finger (Hey, now that's what they should have called a "pinwheel"). (See Figure 10.3.)

[*Editor's Note*: Please, please, use tape to attach the pin to your finger. Otherwise, you might get an infection and sue us.]

A curlometer doesn't spin as readily as a pinwheel. If you blow along its axis, the thing won't turn at all. All your hot air will just glide right past those blades. And if you put the curlometer out into a steady constant wind, it still won't turn, because the wind that pushes the blades clockwise is exactly balanced by the wind that pushes the blades counterclockwise. It's completely useless!

But don't give up yet. It does twirl in certain winds. When the wind vector field varies from point to point, and the wind on one side is stronger than the wind on the other side, then it twirls like a baton at the Marching Band Olympics. Try it—if you just blow on the blades on one side of it, it turns beautifully. It's measuring curl, the tendency of a vector field to twirl things around an axis.

There are some positions you can hold it in where it won't spin at all, and some where it rotates like crazy.

Here's how to use it to measure the curl at a given point: Hold it so that the center of the curlometer, your fingertip, stays at the point you're interested in. Now rotate your hand around till you find the direction for the axis that makes the curlometer whirl as fast as possible. Find the absolute best you can do, twirling-wise, with the center of the curlometer immovable. (If your whole body has become immovable, you might want to stop to see when you got your last tetanus shot.) The curl of the wind vector field **W** is a new vector field curl **W**. The vector field curl **W** tells us all about what that curlometer is doing. The magnitude of curl **W** tells us how fast the curlometer is turning. The direction of curl **W** is the direction in which your finger is pointing.

We left out one tiny detail: You want to find the direction the curlometer is rotating the fastest COUNTERCLOCKWISE, when you look toward your fingertip. Otherwise if you and your brother try this together, one of you will end up upside down shouting "I have it!" while the other will be right side up shouting "No, I have it!" and general confusion will ensue. So agree in advance, counterclockwise.

This counterclockwise direction is easy to remember by using the "right-thumb rule." When you take the thumb of your right hand to be pointing in the same direction as the vector curl **W**, your fingers will be curling in the direction that the vector field is twirling. Here's how to remember this. Take the pin out of your finger and insert it in the thumb of your right hand. Now point your thumb so the curlometer is twirling as rapidly as possible in the direction your fingers are pointing. Your thumb is pointing in the same direction as curl **W**. Okay, we're done, you can take the pin out now.

So say we have a vector field $\mathbf{F}(x, y, x) = P(x, y, z)\mathbf{i} + Q(x, y, z)\mathbf{j} + R(x, y, z)\mathbf{k}$. What's the formula for the curl of **F** at a point (x_0, y_0, z_0)? The curl is a vector with three components, and its formula is a bit complicated.

$$\text{curl } \mathbf{F} = \left(\frac{\partial R}{\partial y} - \frac{\partial Q}{\partial z}\right)\mathbf{i} + \left(\frac{\partial P}{\partial z} - \frac{\partial R}{\partial x}\right)\mathbf{j} + \left(\frac{\partial Q}{\partial x} - \frac{\partial P}{\partial y}\right)\mathbf{k}$$

Luckily, there's a shortcut for remembering this. Treating ∇ as the vector

$$\nabla = \frac{\partial}{\partial x}\mathbf{i} + \frac{\partial}{\partial y}\mathbf{j} + \frac{\partial}{\partial z}\mathbf{k}$$

we can get curl **F** by taking the cross product of ∇ and **F**:

$$\text{curl } \mathbf{F} = \nabla \times \mathbf{F} = \begin{vmatrix} \mathbf{i} & \mathbf{j} & \mathbf{k} \\ \frac{\partial}{\partial x} & \frac{\partial}{\partial y} & \frac{\partial}{\partial z} \\ P & Q & R \end{vmatrix}$$

Example 3 Find the curl of the vector field $\mathbf{F}(x, y, z) = 3x^2z\mathbf{i} + 4xyz\mathbf{j} + yz^2\mathbf{k}$.

Solution

$$\text{curl } \mathbf{F} = \nabla \times \mathbf{F} = \begin{vmatrix} \mathbf{i} & \mathbf{j} & \mathbf{k} \\ \dfrac{\partial}{\partial x} & \dfrac{\partial}{\partial y} & \dfrac{\partial}{\partial z} \\ P & Q & R \end{vmatrix}$$

$$= \left(\frac{\partial R}{\partial y} - \frac{\partial Q}{\partial z} \right)\mathbf{i} + \left(\frac{\partial P}{\partial z} - \frac{\partial R}{\partial x} \right)\mathbf{j} + \left(\frac{\partial Q}{\partial x} - \frac{\partial P}{\partial y} \right)\mathbf{k}$$

$$= (z^2 - 4xy)\mathbf{i} + (3x^2 - 0)\mathbf{j} + (4yz - 0)\mathbf{k}$$

$$= (z^2 - 4xy)\mathbf{i} + 3x^2\mathbf{j} + 4yz\mathbf{k}$$

Hey, whatever twirls your baton.

10.3 Line up for line integrals

While ordinary integrals are evaluated over an interval $[a, b]$, line integrals are evaluated over a curve in space. Line integrals are also called path integrals, and paths are also called curves. Paths that start and end at the same point are called loops or closed curves.

There are several different kinds of line integrals. The first kind involves integrating a function over a curve. You will remember in Section 6.2 that we figured out how to find the arc length of a curve C, described by the parametrization $\mathbf{r}(t) = \langle x(t), y(t), z(t) \rangle$, as t varies from $t = a$ to $t = b$. The idea was to cut the curve up into lots of little pieces, so small that they were almost straight-line segments. The length of each piece was approximately the speed $|\mathbf{r}'(t)|$ times the time interval Δt. Adding all these up, and taking a limit as the size of the time intervals shrunk to 0, gave us the arc length as an integral:

$$\text{Length} = \int_C ds = \int_a^b \sqrt{[x'(t)]^2 + [y'(t)]^2 + [z'(t)]^2} \, dt$$

But now we may want to find some quantity other than the arc length. For example, we can integrate the density function of a curvy piece of wire to get the total mass of the wire. The mass of a little segment of wire is about equal to the product of the length of that segment and the density $\sigma(x, y, z)$ at a spot on the segment, or $\sigma \, ds$. The total mass of the wire is given by

$$\text{Mass} = \int_C \sigma \, ds = \int_a^b \sigma(x(t), y(t), z(t)) \sqrt{[x'(t)]^2 + [y'(t)]^2 + [z'(t)]^2} \, dt$$

Example (Total Mass) The artist Tofu Dancer is designing a necklace in the shape of a wire, strung around a circle with radius 1 foot. Its density at a point (x, y) is given by $\sigma(x, y) = 3 + 2y$ pounds per foot. What is the total mass of the necklace?

Solution We first parametrize the unit circle as $\mathbf{r}(t) = \langle \cos 2\pi t, \sin 2\pi t \rangle$, with $0 \le t \le 1$. Let's calculate ds. Since $x(t) = \cos 2\pi t$ and $y(t) = \sin 2\pi t$ and there is no $z(t)$ around in this problem, we get

$$ds = \sqrt{[2\pi(-\sin 2\pi t)]^2 + [2\pi(\cos 2\pi t)]^2}\, dt = 2\pi\, dt$$

Meanwhile, the density in terms of t becomes

$$\sigma(x, y) = 3 + 2y = 3 + 2\sin 2\pi t$$

So the integral for the total mass becomes

$$
\begin{aligned}
\text{Mass} = \int_C \sigma\, ds &= \int_0^1 [3 + 2\sin 2\pi t] 2\pi\, dt \\
&= \int_0^1 6\pi + 4\pi \sin 2\pi t\, dt \\
&= [6\pi t - 2\cos 2\pi t]_0^1 \\
&= [6\pi - 2\cos 2\pi] - [-2\cos 0] \\
&= 6\pi \\
&\approx 18.85 \text{ pounds}
\end{aligned}
$$

Now that's a lot of necklace.

10.4 Line integrals of vector fields

The work we do in moving something, whether a particle or a piano, is the product of how far we move the thing times how much force we are pushing against as we move it. That's why moving companies charge extra to move a piano into a fifth-floor walk-up.

The work done by a force, like gravity, in moving a particle in a straight line, is the dot product $\mathbf{F} \cdot \mathbf{D}$ of the force vector $\mathbf{F}$ with the displacement vector $\mathbf{D}$. The displacement vector is the vector that points from where the particle began to where the particle ends up. This means that if you lug a piano up four flights, then the work done by gravity is negative, while the amount done by your muscles is positive. If you drop the piano out the fifth-floor window, then the work done by gravity as it falls is positive. The net amount of work done by gravity on the round trip is zero.

Suppose instead that we are not moving a particle along a straight line. Let's face it. How often in life do you move a particle in a straight line? Usually you are zigzagging all over the place, through the hallway, around the corner, up the stairs. So let's suppose that we are moving the particle along a curve. And what about that force vector? What if the force varies depending on where we are? Then we need to represent the force by a vector field, the vectors of which change depending upon where we are on the curve we're moving the particle along.

To compute the work, we think of splitting up the curve into lots of tiny little pieces. As we move along one of the very little pieces, it is almost a straight line. And the force vectors change so little over that little piece that we can assume there is a fixed force vector for that little piece. So we can compute the work in moving the particle along that little piece as the dot product of a fixed force vector and the short displacement vector that approximates our motion. Summing up the approximate work as we move the particle along all the tiny bits of the curve gives us an approximation to the total work of moving the particle along the entire curve. By taking our pieces smaller and smaller, we get a limit which turns the sum into an integral. To calculate these integrals, we parametrize the curve C by a vector valued function $\mathbf{r}(t)$, write everything as a function of t, take dot products, and eventually have an ordinary old integral in t from a to b to evaluate:

$$\text{Work} = \int_C \mathbf{F} \cdot d\mathbf{r} = \int_a^b \mathbf{F}(r(t)) \cdot \mathbf{r}'(t)\, dt$$

Example 1 (Cold Soda in a Hot Tub) Internet retailer SodaEverywhere.com is designing its new soda-dispensing hot tub. The force of the circulating water is described by the vector field $\mathbf{F}(x, y) = x^2\mathbf{i} - 2xy\mathbf{j}$. The bench with the best view of the built-in TV is located at $(0, 1)$. The cooler containing the ice cold cola is located at $(1, 0)$. (See Figure 10.4.) Market research shows that the targeted customer will move in a straight line from the bench to the cooler several dozen times in an afternoon. How much work will the customer need to do each time he moves from the bench to the soda?

Solution We need to work out the line integral for work, $\int_C \mathbf{F} \cdot d\mathbf{r}$. We first come up with a parametrized curve $\mathbf{r}(t)$ which traces out C. This curve, or path, is a straight line, so the curve is given by $\mathbf{r}(t) = (1-t)\mathbf{A} + (t)\mathbf{B}$, where $\mathbf{A}$ and $\mathbf{B}$ are the initial and final position vectors. So

$$\mathbf{r}(t) = (1-t)\langle 0, 1\rangle + (t)\langle 1, 0\rangle = \langle t, 1-t\rangle = (t)\mathbf{i} + (1-t)\mathbf{j}$$

with $0 \le t \le 1$. Now we calculate $\mathbf{F}(t)$, which is the same as $\mathbf{F}(\mathbf{r}(t))$.

$$\mathbf{F}(\mathbf{r}(t)) = \mathbf{F}(t, 1-t) = t^2\mathbf{i} - 2t(1-t)\mathbf{j} = t^2\mathbf{i} + (2t^2 - 2t)\mathbf{j}$$

(0, 1)

(1, 0)

Figure 10.4 A vector field drinking problem.

Next we calculate

$$\mathbf{r}'(t) = (1)\mathbf{i} + (-1)\mathbf{j} = \mathbf{i} - \mathbf{j}$$

The dot product is then

$$\mathbf{F}(\mathbf{r}(t)) \cdot \mathbf{r}'(t) = [(t^2)\mathbf{i} + (2t^2 - 2t)\mathbf{j}] \cdot [\mathbf{i} - \mathbf{j}]$$
$$= t^2 - (2t^2 - 2t)$$
$$= 2t - t^2$$

The path $\mathbf{r}(t)$ is traversed as t takes values $0 \le t \le 1$. So the total work done per trip is given by the familiar integral

$$\int_0^1 2t - t^2 \, dt = \left[t^2 - \frac{t^3}{3} \right]_0^1 = \frac{2}{3}$$

The amount of work done by the circulating water is positive, so the customer is actually pushed toward those six-packs by the churning liquid. (When the force and the motion line up and give a positive inner product, the work done by the force is positive.) The company makes more on the cola than on the hot tub, so the engineer who designed that tub got a lot of stock options.

WHAT'S THAT $P \, dx + Q \, dy$ THING?

Often we will see line integrals written like

$$\int_C P \, dx + Q \, dy$$

This kind of integral comes up when we are calculating the work done in moving a particle along a path. The expression $P \, dx + Q \, dy$ breaks down the integral into its x- and y-components. When we have a vector field

$$\mathbf{F} = P(x, y)\mathbf{i} + Q(x, y)\mathbf{j}$$

and a parametrized curve

$$\mathbf{r}(t) = x(t)\mathbf{i} + y(t)\mathbf{j} \qquad a \le t \le b$$

then

$$\mathbf{r}'(t) = \frac{dx}{dt}\mathbf{i} + \frac{dy}{dt}\mathbf{j}$$

The dot product of $\mathbf{F}$ and $\mathbf{r}'(t)$ is given by

$$\mathbf{F}(r(t)) \cdot \mathbf{r}'(t) = \left(P\frac{dx}{dt} + Q\frac{dy}{dt} \right)$$

So the work

$$W = \int_C \mathbf{F}(r(t)) \cdot \mathbf{r}'(t) \, dt = \int_C \left(P\frac{dx}{dt} + Q\frac{dy}{dt} \right) dt = \int_C P \, dx + Q \, dy$$

The $P\,dx + Q\,dy$ form is particularly useful when the line integral is over a graph $y = f(x)$. This is just a parametrized curve with $x = t$ and $y = f(t)$. We then can calculate a line integral as a standard integral over an interval $a \le x \le b$.

Example 2   Calculate the line integral of $\mathbf{F} = 2x\mathbf{i} - 6x^2\mathbf{j}$ for the curve C which is the graph of $y = x^2 - x$ over $1 \le x \le 2$.

Solution   We substitute $y = x^2 - x$ and $dy = (2x - 1)\,dx$ to get

$$\int_C P\,dx + Q\,dy = \int_1^2 (2x)\,dx + (-6x^2)(2x - 1)\,dx$$

This is now a completely standard integral.

$$\int_1^2 2x - 12x^3 + 6x^2\,dx = [x^2 - 3x^4 + 2x^3]_1^2 = -28$$

10.5 Conservative vector fields

Sometimes a line integral of a vector field along a curve can be calculated more easily. This happens when the vector field is the gradient of a function, a common situation when studying electric and gravitational forces.

> The gradient of a function f on a region R is a *conservative* vector field.

The situation here is similar to a situation we have seen before, namely, the *fundamental theorem of calculus*. It told us that when $g(x) = G'(x)$ then $\int_a^b g(x)\,dx = G(b) - G(a)$, and that made calculating integrals a lot easier. Similarly, when C is a curve in a region R going from a point given by $\mathbf{A}$ to a point given by $\mathbf{B}$, and $\mathbf{F} = \nabla f$ in R there's an easy way to calculate $\int_C \mathbf{F} \cdot d r$.

> When $\mathbf{F} = \nabla f$,
>
> $$\int_C \mathbf{F} \cdot d r = \int_C \nabla f \cdot d r = f(\mathbf{B}) - f(\mathbf{A})$$

We can think of this as a vector version of the fundamental theorem of calculus. Let's see how easy it can be to calculate with these conservative fields:

Example 1   Check that $\mathbf{F} = (2x+3y)\mathbf{i}+(3x-2y)\mathbf{j}$ is conservative, and compute the line integral $\displaystyle\int_C \mathbf{F}\cdot d\mathbf{r}$ along the path which follows the quarter of a unit radius circle going counterclockwise from (1, 0) to (0, 1).

Solution   We check that $\mathbf{F}$ is conservative by finding a function $f(x, y)$ with $\mathbf{F} = \nabla f$. Here it is:

$$f(x, y) = x^2 + 3xy - y^2$$

This function has gradient equal to $\mathbf{F}$ at all (x, y).

But how did we find this function? Well, since $\nabla f = \mathbf{F}$, we must have that

$$\frac{\partial f}{\partial x} = 2x + 3y$$

Integrating with respect to the variable x gives $f(x, y) = x^2 + 3xy + g(y)$, where $g(y)$ is any function depending on y but not on x, and therefore having 0 as its partial derivative with respect at x. Differentiating this with respect to y gives

$$\frac{\partial f}{\partial y} = 3x + g'(y)$$

Setting this equal to the second component $(3x - 2y)$ of $\mathbf{F}$, we see that

$$g'(y) = -2y$$

and so

$$g(y) = -y^2 + K$$

We can take the constant K to be zero since we're calculating a definite integral (any choice gives the same result). So we see that

$$f(x, y) = x^2 + 3xy - y^2$$

is a function such that $\nabla f = \mathbf{F}$. Note that $f(x, y) = x^2 + 3xy - y^2 + K$ is another function with $\nabla f = \mathbf{F}$, for any constant K.

This means **F** is conservative, so it's easy to find the line integral. We don't need to parametrize a curve or take any dot products. We just calculate

$$\int_C \mathbf{F} \cdot \mathbf{d}r = \int_C \nabla f \cdot \mathbf{d}r = f(0, 1) - f(1, 0) = (-1) - (1) = -2$$

When we go around a loop, the starting point **A** and the end point **B** are the same, **A** = **B**. This means that the line integral of a conservative vector field along a loop is $f(\mathbf{B}) - f(\mathbf{A}) = f(\mathbf{A}) - f(\mathbf{A}) = 0$. A particularly easy calculation.

KEY PROPERTIES OF CONSERVATIVE VECTOR FIELDS

1. The line integral of a conservative vector field along a curve can be calculated by taking the difference of its potential function at the two endpoints of the curve.

2. The line integral of a vector field which is conservative in a region does not depend on which parametrized curve in the region is used, just on the curve's starting and ending points. It is *independent of path.*

3. The line integral of a conservative vector field in a region around a loop in that region equals zero. (Conservative vector fields go around in a circle and do nothing. By contrast, nonconservative vector fields can go around in a circle and run up a deficit.)

So how do we tell if a vector field is conservative? First let's see how to tell that it's NOT conservative. If a vector field $\mathbf{F} = P\mathbf{i} + Q\mathbf{j}$ is conservative, then $\mathbf{F} = \nabla f$ for some function $f(x, y)$, so $P = \partial f/\partial x$ and $Q = \partial f/\partial y$. If we take the partial of P with respect to y, we find

$$\frac{\partial P}{\partial y} = \frac{\partial}{\partial y}\left(\frac{\partial f}{\partial x}\right) = \frac{\partial^2 f}{\partial y\, \partial x} = \frac{\partial^2 f}{\partial x\, \partial y} = \frac{\partial}{\partial x}\left(\frac{\partial f}{\partial y}\right) = \frac{\partial Q}{\partial x}$$

Consequently,

A conservative vector field always satisfies

$$\frac{\partial P}{\partial y} = \frac{\partial Q}{\partial x}$$

If these partial derivatives are not equal or don't exist somewhere in a region, then the vector field is not conservative in the region.

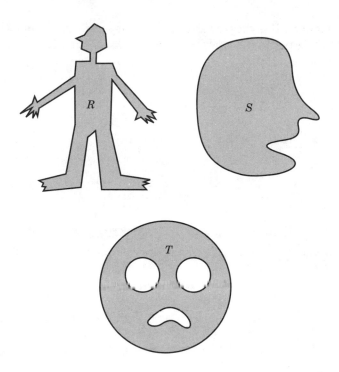

Figure 10.5 A region with no holes (like R and S, but not like T) is called *simply connected*.

Example 2   Is $\mathbf{F} = x^2 y\mathbf{i} + x^3\mathbf{j}$ a conservative vector field?

Solution $\partial P/\partial y = x^2$ and $\partial Q/\partial x = 3x^2$ so $\partial P/\partial y \neq \partial Q/\partial x$. Therefore this vector field is NOT conservative. (It's also not liberal, more of a middle-of-the-road type.)

What if $\partial P/\partial y = \partial Q/\partial x$? Can we then conclude the vector field is conservative? Can we find an $f(x, y)$ with $\nabla f = \mathbf{F}$? Not always. We need another condition to be sure that we can find a potential function $f(x, y)$ with $\nabla f = \mathbf{F}$. This extra condition is that $\mathbf{F}$ is defined on a domain with no holes in it, a region which is called *simply connected*. Regions R and S, but not the unhappy region T, are simply connected in Figure 10.5.

> If $\partial P/\partial y = \partial Q/\partial x$ on a *simply connected* region R, then $\mathbf{F} = \nabla f$ for some function $f(x, y)$ on R, and so $\mathbf{F}$ is conservative on R.

Common Mistake $\partial P/\partial y = \partial Q/\partial x$ must be true everywhere in a simply connected region for $\mathbf{F} = P\mathbf{i} + Q\mathbf{j}$ to be conservative. It's not enough for it to be true along a path or loop. It must hold everywhere inside the loop before we can assert that there is an f with $\nabla f = \mathbf{F}$ and that the line integral of $\mathbf{F}$ over the loop is zero.

Example 3   Take $\mathbf{F}$ to be the vector field

$$\mathbf{F} = \frac{y}{x^2 + y^2}\mathbf{i} + \frac{-x}{x^2 + y^2}\mathbf{j}$$

Compute $\oint_C \mathbf{F} \cdot d\mathbf{r}$ along the path C which goes counterclockwise once around the unit circle, starting and ending at (1, 0).

Note   We have written the integral symbol $\oint$ with a little circle through it. This indicates that we are doing a line integral around a loop, a path that starts and ends at the same point.

Solution   Let's check to see if $\mathbf{F}$ is conservative. If it is, the problem is easy, because the path we are integrating over is a loop, starting and ending at the same point, and the line integral of a conservative vector field over a loop equals zero.
So let's calculate. By the quotient rule

$$\frac{\partial P}{\partial y} = \frac{(x^2 + y^2)(1) - (2y)(y)}{(x^2 + y^2)^2} = \frac{x^2 - y^2}{(x^2 + y^2)^2}$$

and

$$\frac{\partial Q}{\partial x} = \frac{(x^2 + y^2)(-1) - (2x)(-x)}{(x^2 + y^2)^2} = \frac{x^2 - y^2}{(x^2 + y^2)^2}$$

So these two are equal and it is tempting to say that $\mathbf{F}$ is conservative. But that would be WRONG. The problem is that $\mathbf{F}$, $\partial P/\partial y$ and $\partial Q/\partial x$ are not defined at the point (0, 0), where there is a division by zero. The region where they are equal excludes (0, 0), and is NOT simply connected. And in fact $\mathbf{F}$ is NOT conservative on a region containing the whole loop. There is no function $f(x, y)$ whose gradient equals $\mathbf{F}$ and whose domain contains the whole curve C.
Let's calculate what the line integral actually is. We'll have to parametrize the curve and calculate. So take

$$\mathbf{r}(t) = \cos t\mathbf{i} + \sin t\mathbf{j} \qquad 0 \le t \le 2\pi$$

to be a parametrization of the unit circle. Now $x = \cos t$ and $y = \sin t$ along C, so that $x^2 + y^2 = 1$. Then $\mathbf{F}$ simplifies to

$$\mathbf{F} = \sin t\mathbf{i} - \cos t\mathbf{j}$$

and

$$\mathbf{r}'(t) = -\sin t\mathbf{i} + \cos t\mathbf{j}$$

Then

$$\mathbf{F} \cdot \mathbf{r}'(t) = -\sin^2 t - \cos^2 t = -1$$

So

$$\oint_C \mathbf{F} \cdot d\mathbf{r} - \int_0^{2\pi} -1 \, dt = \quad 2\pi$$

This is not zero, as it would be if $\mathbf{F}$ were conservative. Whew, good thing we were careful!

10.6 Green's theorem

If the difference $\partial Q/\partial x - \partial P/\partial y$ is not zero, then $\mathbf{F}$ is not conservative, and the line integral does depend on the path taken. It may turn out not to be zero along a loop. Green's theorem tells us how to calculate the line integral of any vector field around a loop. We have to go around the loop counterclockwise to make this result work. The region D must be on our left as we walk around it along its boundary curve C. (See Figure 10.6). Counterclockwise is referred to as "positively oriented."

Theorem (Green's Theorem in the Plane)   *If C is a positively oriented loop which encloses a region D and $\mathbf{F} = P\mathbf{i} + Q\mathbf{j}$ is a vector field on D, then*

$$\oint_C P \, dx + Q \, dy = \iint_D \left(\frac{\partial Q}{\partial x} - \frac{\partial P}{\partial y} \right) dx \, dy$$

In the case of a conservative vector field, both sides of the equation are zero.

This is a pretty amazing result when you think about it. A double integral over an entire region is equal to a line integral that takes place only on

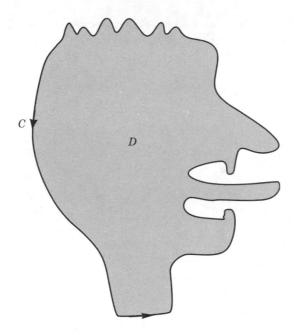

Figure 10.6 Green's theorem states that a line integral over the curve C equals an integral over the region D.

the boundary of that region. It's sort of like judging everything in a book by looking only at the boundary—with vector fields you really can judge a book by its cover.

Technical Point   In Green's theorem and the other integral theorems, the vector field **F** and the other functions that appear, $\partial Q/\partial x$ and $\partial P/\partial y$, must be continuous, or we can run into the type of problem we saw in Example 3 in Section 10.5.

Example 1   Calculate the line integral $\oint_C -y\,dx + x\,dy$, where C is a circle of radius R in the plane, centered at $(0, 0)$, traversed counterclockwise.

Solution   Applying Green's theorem, we see that this equals an integral over the region inside the circle, which we call D. In this particular case, $P = -y$ and $Q = x$, so $(\partial Q/\partial x - \partial P/\partial y) = 1 - (-1) = 2$ and

$$\oint_C -y\,dx + x\,dy = \iint_D 2\,dx\,dy = 2\,(\text{area}\,(D)) = 2\pi R^2$$

So this line integral equals twice the area of the disk. Doesn't matter where the disk is. Even for a curve that's not round, the line integral of $-y \, dx + x \, dy$ along it gives twice the area enclosed by the curve.

Example 2   Calculate the line integral

$$\oint_C (x^3 \sin x - 5y) \, dx + (4x + e^{y^2}) \, dy$$

where C is the circle of radius 2 in the plane, centered at $(0, 0)$, traversed counterclockwise.

   Solution   While we could parametrize C and calculate everything out, Green's theorem gives a much easier way to proceed. We have $P = x^3 \sin x - 5y$ and $Q = 4x + e^{y^2}$. We calculate

$$\frac{\partial Q}{\partial x} = 4 \qquad \text{and} \qquad \frac{\partial P}{\partial y} = -5$$

So

$$\left(\frac{\partial Q}{\partial x} - \frac{\partial P}{\partial y} \right) = 4 + 5 = 9$$

and

$$\oint_C (x^3 \sin x - 5y) \, dx + (4x + e^{y^2}) \, dy = \iint_D 9 \, dx \, dy$$

The integral of 9 over D, the region enclosed by C, is just 9 times the area of D, or $9\pi r^2$ with $r = 2$.

$$9\pi(2)^2 = 36\pi$$

Wow, was that ever easy! Whatever battens down your hatches.

10.7 Integrating the divergence; the divergence theorem

Green's theorem has a cousin which tells what happens when we integrate the divergence of a function, called appropriately enough the divergence theorem. There are two versions, one for the plane and one for space.

Remember that divergence measures the net inflow or outflow of a vector field at a point. When we integrate divergence over a region, we get the net amount of stuff that is being created or destroyed in the whole region. As you might expect, this is equal to the amount flowing in or out of the region across its boundary, and that is just what the divergence theorem says.

Example 1 (A Toxic Example)   Toxic waste is swirling around in a shallow superfund site surrounded by a fence. Waste is being introduced at some points (leaking drums). These points are called "sources" and have positive divergence. At other points called "sinks," the waste gets sucked away and has negative divergence. (In the old days toxic waste was flushed down the sink, hence the name.) The toxic waste sloshes around, giving rise to a velocity vector field. At each point in the site the velocity vector field measures the direction and speed at which the liquid (or sludge or PCBs) at that point is moving. A certain amount flows out of the boundary of the toxic waste site and into the neighboring homes. The question everyone wants the answer to is "How will it affect property values?" or in other words, "How much is flowing out?" The divergence theorem tells us the answer.

Solution   By adding up (integrating) the total divergence of the velocity vector field, we get the total flow through the fence.

Example 2 (The Party Line Integral*)   A huge rambunctious party is going on inside a large room. Numerous people mill about, coming and going. The motion of the party goers forms a party velocity vector field. (Sociologists analyze these fields by studying the patterns of drool and spilled beer left on the floor.) Certain people act like magnets. Other party goers are attracted to them, sometimes because of their scintillating conversation, and sometimes because of how they look in a sleeveless T-shirt. They are surrounded by admirers. Though they don't often think of themselves that way, they are what we call "sinks." Other party goers are repellers, sometimes because they can only talk about their most recent root canal and sometimes because of how they look in a sleeveless T-shirt. Revelers move away from them as quickly as politeness allows, and often quicker. They are known as "sources." Most people are neither sources nor sinks, but just go with the flow.

The divergence theorem tells us the relationship between the sources and sinks and the number of people entering and leaving the room. It says that if we sum up, by integrating, the divergence of the party vector field, then

---

*Thanks to J. Kasdan for this one.

we get the total number of people entering or leaving the room, either by sneaking past the bouncers at the door or by falling off the balcony into the pool.

CALCULATING WITH THE DIVERGENCE THEOREM

Let D be a region in the plane whose boundary is a curve C. In the example above, that region was the floor of the room. Let $\mathbf{F}$ be a vector field defined over the region D. To calculate with the divergence theorem, we need to measure how much stuff flows across the boundary of a region. This is given by integrating the flow of the vector field $\mathbf{F}$ across C, or calculating the "flux" (Latin for flow—and we thought our four years of Latin were wasted.)

The amount of stuff being pushed across the curve is found by calculating what gets pushed across a tiny bit of curve, adding up for each little bit, and taking the limit to get an integral. The amount of stuff crossing a little curve piece with length ds is given by taking the part of $\mathbf{F}$ which is perpendicular to that bit of curve, since the part of $\mathbf{F}$ that is parallel to the curve doesn't push anything across it, and multiplying by the length ds. The normal part of $\mathbf{F}$, the part of $\mathbf{F}$ which is perpendicular to C, is obtained by taking the dot product of $\mathbf{F}$ with a vector $\mathbf{n}$ of length 1 that's normal, or perpendicular, to C. (See Figure 10.7.)

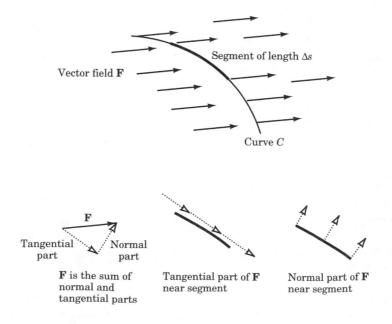

Figure 10.7 The amount of flow of a vector field moving across a segment of C is the product of the length of the segment with the length of the normal part of $\mathbf{F}$.

A unit length normal vector is usually called $\mathbf{n}$. To calculate $\mathbf{n}$ for a curve $\mathbf{r}(t) = \langle x(t), y(t) \rangle$ in the plane, we use the formula

$$\mathbf{n} = \frac{y'(t)}{|\mathbf{r}'(t)|}\mathbf{i} + \frac{-x'(t)}{|\mathbf{r}'(t)|}\mathbf{j}$$

where $|\mathbf{r}'(t)| = \sqrt{(x'(t))^2 + (y'(t))^2}$. This gives a vector of length 1 whose dot product with the tangent vector $x'(t)\mathbf{i} + y'(t)\mathbf{j}$ is zero.

The flux is then given by

$$\text{Flux} = \oint_C \mathbf{F} \cdot \mathbf{n} \, ds$$

The divergence theorem says that this integral that gives the flux is in fact equal to the integral of the divergence of $\mathbf{F}$, denoted $\nabla \cdot \mathbf{F}$, over D, the region enclosed by C. The curve C must be parametrized so that the region D is on the left as we walk along C. The divergence theorem is very similar to Green's theorem, but the things being integrated are a little different. With Green's theorem the line integral integrates the tangential part of the vector field over the boundary, while with the divergence theorem it integrates the perpendicular part over the boundary.

Here's the actual theorem:

Theorem (Divergence Theorem in the Plane)

$$\text{Flux} = \oint_C \mathbf{F} \cdot \mathbf{n} \, ds = \iint_D \text{div} \, \mathbf{F} \, dx \, dy$$

(Continuity is required of $\mathbf{F}$ and div $\mathbf{F}$ in D.)

Example 3   Calculate the flux of the vector field $\mathbf{F} = x\mathbf{i} + y\mathbf{j}$ across the circle of radius 3 centered at $(0, 0)$.

Solution   The divergence theorem makes this easy.

$$\text{div} \, \mathbf{F} = \nabla \cdot \mathbf{F} = \frac{\partial P}{\partial x} + \frac{\partial Q}{\partial y} = 1 + 1 = 2$$

The theorem tells us the flux we want to know is given by

$$\text{Flux} = \iint_D \text{div} \, \mathbf{F} \, dx \, dy = \iint_D 2 \, dx \, dy = 2(\text{area} \, (D)) = 2\pi(3)^2 = 18\pi$$

So the amount of stuff that's flowing out of D is twice the area of D, in this case 18π. If this is the vector field that models your local superfund site, either sell your house or start organizing a class action lawsuit. If this is the vector field that models your party, then it looks like quite a few people are falling off the balcony into the pool.

10.8 Surface integrals

We would like to be able to calculate integrals over surfaces. Why? Perhaps we're in the business of coating Edam cheese with those red wax covers and need to know how much wax to buy for different shapes of cheese. Or maybe we need to calculate the mass of a surface, where the density varies over the surface. If we are going to have any hope of computing integrals on a surface S sitting in space, we need to describe S with equations.

We have already seen one way to describe such a surface, which works when it's the graph of an equation $z = f(x, y)$. But this isn't the only way to describe a surface. There is another option.

Just as we had parametrized curves $\mathbf{r}(t)$, we have parametrized surfaces $\mathbf{r}(u, v)$. This time there are TWO variables in the domain, u and v, because we're looking at a two-dimensional object. Just as a longitude and a latitude are enough to tell us where we are on the surface of the Earth, with two coordinates we can tell where we are on a surface. The equations describing a parametric surface look like

$$\mathbf{r}(u, v) = x(u, v)\mathbf{i} + y(u, v)\mathbf{j} + z(u, v)\mathbf{k}$$

For each choice of u and v, $\mathbf{r}(u, v)$ is a point in space. So, for example, if we pick $u = 2$ and $v = 5$, then $\mathbf{r}(u, v)$ is the point in space whose x-coordinate is given by $x(2, 5)$, whose y-coordinate is given by $y(2, 5)$, and whose z-coordinate is given by $z(2, 5)$.

A special case is the situation we've looked at before, when a surface is given as the graph of $z = f(x, y)$. In vector form, this becomes

$$\mathbf{r}(x, y) = x\mathbf{i} + y\mathbf{j} + f(x, y)\mathbf{k}$$

The z-coordinate here is a function of the x- and y-coordinates. For each choice of x and y there is only one z-value. This kind of formula is good for modeling things like roofs or circus tents, where there is at most one point of the surface overhead.

PARAMETRIZED SURFACES AND THEIR AREAS

The integral for the area of a parametric surface S is found, as usual, by a limiting process. To get the formula, we first need the area of a little

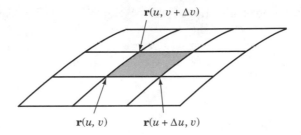

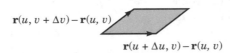

Figure 10.8 The area of the little parallelogram on the surface is about $|(\mathbf{r}(u + \Delta u, v) - \mathbf{r}(u, v)) \times (\mathbf{r}(u, v + \Delta v) - \mathbf{r}(u, v))|$.

parallelogram whose sides are given by vectors $\mathbf{A}$ and $\mathbf{B}$. As we saw in Section 5.5,

$$\text{Area (parallelogram)} = |\mathbf{A} \times \mathbf{B}|$$

Now suppose that $\mathbf{r}(u, v)$ describes the surface. A little parallelogram on the surface formed by the edges from $\mathbf{r}(u, v)$ to $\mathbf{r}(u + \Delta u, v)$ and from $\mathbf{r}(u, v)$ to $\mathbf{r}(u, v + \Delta v)$, as in Figure 10.8, has area approximately given by

$$\text{Area (parallelogram)} = |(\mathbf{r}(u + \Delta u, v) - \mathbf{r}(u, v)) \times (\mathbf{r}(u, v + \Delta v) - \mathbf{r}(u, v))|$$

This little area equals

$$\left| \left(\frac{\mathbf{r}(u + \Delta u, v) - \mathbf{r}(u, v)}{\Delta u} \right) \times \left(\frac{\mathbf{r}(u, v + \Delta v) - \mathbf{r}(u, v)}{\Delta v} \right) \right| \Delta u \, \Delta v$$

In the limit this becomes

$$dS = |\mathbf{r}_u \times \mathbf{r}_v| \, du \, dv$$

where $\mathbf{r}_u = \partial \mathbf{r}/\partial u$ and $\mathbf{r}_v = \partial \mathbf{r}/\partial v$. We call dS the *element of surface area* and think of it as the area of a tiny bit of surface.

Adding all these tiny areas up (by integrating) gives the area of the entire surface.

$$\text{Area } (S) = \iint_S dS = \iint_S |\mathbf{r}_u \times \mathbf{r}_v| \, du \, dv$$

A particularly simple situation is when S is given as a graph of $z = f(x, y)$, where (x, y) lie in a region D in the xy-plane. This has a parametrization, where $x = u, y = v,$ and $z = f(x, y)$. So $\mathbf{r}(x, y) = \langle x, y, f(x, y) \rangle$. Then, $\mathbf{r}_u = \mathbf{r}_x = \langle 1, 0, \partial f/\partial x \rangle$, and $\mathbf{r}_v = \mathbf{r}_y = \langle 0, 1, \partial f/\partial y \rangle$. So $\mathbf{r}_u \times \mathbf{r}_v = \langle -\partial f/\partial x, -\partial f/\partial y, 1 \rangle$. The area of a tiny piece dS of S is then given by

$$dS = |\mathbf{r}_u \times \mathbf{r}_v| \, du \, dv = \sqrt{1 + \left(\frac{\partial f}{\partial x}\right)^2 + \left(\frac{\partial f}{\partial y}\right)^2} \, dx \, dy$$

The area of the surface comes from integrating this.

The area of a graph $z = f(x, y)$ over a region D in the xy-plane:

$$\text{Area} = \iint_D \sqrt{1 + \left(\frac{\partial f}{\partial x}\right)^2 + \left(\frac{\partial f}{\partial y}\right)^2} \, dx \, dy$$

Example 1 (Area of a Cylinder) Find a parametric representation of the cylinder of radius 3 around the z-axis between $z = 0$ and $z = 10$, and calculate its area.

Solution We parametrize the circle of radius 3 by

$$\mathbf{C}(\theta) = \langle 3 \cos \theta, 3 \sin \theta \rangle \qquad 0 \le \theta \le 2\pi$$

A point on the cylinder is specified by using θ to give a point on the circle, and adding a z-value to tell how high the point is. So a parametric representation of the cylinder is

$$\mathbf{r}(\theta, z) = \langle 3 \cos \theta, 3 \sin \theta, z \rangle \qquad 0 \le \theta \le 2\pi, 0 \le z \le 10$$

In vector form, this can be written as

$$\mathbf{r}(u, v) = 3\cos u\mathbf{i} + 3\sin u\mathbf{j} + v\mathbf{k} \qquad 0 \le u \le 2\pi, 0 \le v \le 10$$

To calculate the area of S, we calculate $\mathbf{r}_u \times \mathbf{r}_v$.

$$\mathbf{r}_u = -3\sin u\mathbf{i} + 3\cos u\mathbf{j} + 0\mathbf{k}$$

and

$$\mathbf{r}_v = 0\mathbf{i} + 0\mathbf{j} + 1\mathbf{k}$$

Then

$$\mathbf{r}_u \times \mathbf{r}_v = 3\cos u\mathbf{i} + 3\sin u\mathbf{j}$$

and

$$|\mathbf{r}_u \times \mathbf{r}_v| = \sqrt{(3\cos u)^2 + (3\sin u)^2} = 3$$

So

$$\text{Area }(S) = \iint_D 3\, du\, dv = \int_0^{10}\int_0^{2\pi} 3\, du\, dv = \int_0^{10}\left[3u\right]_0^{2\pi} dv$$

$$= \int_0^{10} 6\pi\, dv = \left[6\pi v\right]_0^{10} = 60\pi$$

This is equal to the length of the circle, 6π, times the height 10, which is just as well, since that's another way we get the surface area of a cylinder.

INTEGRATING FUNCTIONS OVER A SURFACE IN SPACE

As we mentioned previously, there are many situations in which we want to integrate a function over a surface. For instance, we might want to calculate the mass of a surface whose density, or mass per unit area, varies over the surface. This occurs with those hollow chocolate bunnies that people eat around Easter. The chocolate is not uniformly distributed over the bunny, but we would want to know the total mass of the chocolate we are eating. We can find the mass by integrating the density function over the surface of the bunny.

A surface S whose density is given by a function $\sigma(x, y, z)$ has total mass given by adding up the density of little bits of the surface times the areas of those little bits. So the total mass is given by

$$\text{Total mass} = \iint_S \sigma(x, y, z)\, dS$$

How we calculate this depends on how the surface S is represented. If S is a parametrized surface $\mathbf{r}(u, v) = \langle x(u, v), y(u, v), z(u, v) \rangle$, then we see that $dS = |\mathbf{r}_u \times \mathbf{r}_v| \, du \, dv$ and the formula for integrating the function $\sigma(x, y, z)$ over the surface is

$$\iint_S \sigma(x(u, v), y(u, v), z(u, v)) \, |\mathbf{r}_u \times \mathbf{r}_v| \, du \, dv$$

Example 2 (Calculus of Roofing)   The architect Svenn Diagram has planned a soaring roof for the new math building. The roof is made of a cutting-edge space-age material of varying density. The roof is described by the graph of the function $z = 10 + 2x + 2y$ over the rectangle $0 \le x \le 20$, $0 \le y \le 15$ (in meters). The density function of the roof at (x, y, z) is given by $\sigma(x, y, z) = (100 - z)$ kilograms per square meter. If the total mass of the roof is greater than 60,000 kilograms, the building collapses on its occupants, and there is no more math at the university. If the total mass of the roof is less than 40,000 kilograms, the roof blows away on the first windy day. Determine the mass of the roof and whether Svenn will ever get another job. (For this problem, assume that people will be upset if the roof collapses on the math department.)

Solution   Since this surface is the graph of $f(x, y) = 10 + 2x + 2y$, we calculate the mass to be

$$\text{Mass} = \iint_S \sigma(x, y, 10 + 2x + 2y) \sqrt{1 + \left(\frac{\partial f}{\partial x}\right)^2 + \left(\frac{\partial f}{\partial y}\right)^2} \, dx \, dy$$

$$= \int_0^{15} \int_0^{20} [100 - (10 + 2x + 2y)] \sqrt{1 + (2)^2 + (2)^2} \, dx \, dy$$

$$= \int_0^{15} \int_0^{20} 270 - 6x - 6y \, dx \, dy$$

$$= \int_0^{15} \left[270x - 3x^2 - 6xy \right]_{x=0}^{x=20} dy$$

$$= \int_0^{15} 4200 - 120y \, dy$$

$$= \left[4200y - 60y^2 \right]_{y=0}^{y=15}$$

$$= 49{,}500 \text{ kilograms}$$

So the mass is within the right bounds, and the roof will stay put. Unfortunately, Svenn didn't bother to put any bathrooms in the building, so the beautiful new structure is surrounded by Port-a-Pottys. Looks like Svenn might have trouble finding his next commission.

FLUX OF A VECTOR FIELD ACROSS A SURFACE IN SPACE

We can compute the flux of a vector field $\mathbf{F}$ across a surface S just like we did for a curve in the plane.

The *surface integral*, or *flux* of $\mathbf{F}$ across S, is given by

$$\text{Flux} = \iint_S \mathbf{F} \cdot \mathbf{n} \, dS$$

where $\mathbf{n}$ is a unit length normal vector field to S. The flux is measuring the flow of $\mathbf{F}$ across the surface S.

Let's take a closer look at $\mathbf{n} \, dS$ when we have a parametrization of S over a region D in the plane. Since $\mathbf{r}_u \times \mathbf{r}_v$ is a normal vector for the surface, it's equal to a unit normal vector $\mathbf{n}$ times its length $|\mathbf{r}_u \times \mathbf{r}_v|$,

$$\mathbf{r}_u \times \mathbf{r}_v = |\mathbf{r}_u \times \mathbf{r}_v|\mathbf{n}$$

So

$$\mathbf{n} = \frac{\mathbf{r}_u \times \mathbf{r}_v}{|\mathbf{r}_u \times \mathbf{r}_v|}$$

Since $dS = |\mathbf{r}_u \times \mathbf{r}_v| \, du \, dv$, we have that

$$\mathbf{n} \, dS = (\mathbf{r}_u \times \mathbf{r}_v) \, du \, dv$$

So we know that

$$\text{Flux} = \iint_S \mathbf{F} \cdot \mathbf{n} \, dS = \iint_D \mathbf{F} \cdot (\mathbf{r}_u \times \mathbf{r}_v) \, du \, dv$$

The vector $(\mathbf{r}_u \times \mathbf{r}_v)$ is straightforward to calculate.

Example 3 (Bee Flux)   A beehive lies inside a chickenwire cage described by the equation $x^2 + y^2 + z^2 = 1$. The beehive is disturbed as we hike by, and the velocity of the emerging bees is given by the vector field $\mathbf{F}(x, y, z) = x\mathbf{i} + y\mathbf{j} + z\mathbf{k}$. The chickenwire holes are too big to keep the bees in, and the flux of $\mathbf{F}$ over the chickenwire surface measures how many bees are flying across the chickenwire, out of the cage. Calculate this flux.

Solution We'll need to parametrize the sphere, and quickly. We can do this with spherical coordinates (ρ, ϕ, θ). Remember, from Section 9.6, $x = \rho \sin \phi \cos \theta, y = \rho \cos \phi \cos \theta$, and $z = \rho \cos \phi$. We set $\rho = 1$ since we are on a sphere of radius 1. Then the surface is given by

$$\mathbf{r}(\phi, \theta) = (\sin \phi \cos \theta)\mathbf{i} + (\sin \phi \sin \theta)\mathbf{j} + (\cos \phi)\mathbf{k} \qquad 0 \le \phi \le \pi, \ 0 \le \theta \le 2\pi$$

In this case, ϕ and θ are playing the roles of u and v, respectively. We calculate

$$\mathbf{r}_\phi \times \mathbf{r}_\theta = \begin{vmatrix} \mathbf{i} & \mathbf{j} & \mathbf{k} \\ \cos \phi \cos \theta & \cos \phi \sin \theta & -\sin \phi \\ -\sin \phi \sin \theta & \sin \phi \cos \theta & 0 \end{vmatrix}$$

$$= (\sin^2 \phi \cos \theta)\mathbf{i} + (\sin^2 \phi \sin \theta)\mathbf{j} + (\sin \phi \cos \phi)\mathbf{k}$$

Also

$$\mathbf{F}(\mathbf{r}(\phi, \theta)) = (\sin \phi \cos \theta)\mathbf{i} + (\sin \phi \sin \theta)\mathbf{j} + (\cos \phi)\mathbf{k}$$

The dot product is:

$$\begin{aligned} \mathbf{F}(\mathbf{r}(\phi, \theta)) \cdot (\mathbf{r}_\phi \times \mathbf{r}_\theta) &= \sin^3 \phi \cos^2 \theta + \sin^3 \phi \sin^2 \theta + \cos^2 \phi \sin \phi \\ &= \sin^3 \phi + \cos^2 \phi \sin \phi \\ &= (\sin^2 \phi + \cos^2 \phi) \sin \phi \\ &= \sin \phi \end{aligned}$$

So we have

$$\begin{aligned} \text{Flux} &= \int_0^{2\pi} \int_0^\pi \sin \phi \, d\phi \, d\theta \\ &= \int_0^{2\pi} \left[-\cos \phi \right]_0^\pi d\theta = \int_0^{2\pi} 2 \, d\theta \\ &= 4\pi \end{aligned}$$

That's about 12 bees per second. Should get you moving up the trail.

THE DIVERGENCE THEOREM IN 3-SPACE

The divergence theorem for 3-space says that the total divergence of a vector field in a solid region V in space equals the total flow across the boundary surface S of the region.

Theorem (Divergence Theorem in Space)

$$\text{Flux} = \iint\limits_{S} \mathbf{F} \cdot \mathbf{n}\, dS = \iiint\limits_{V} \text{div } \mathbf{F}\, dV$$

($\mathbf{F}$ and div $\mathbf{F}$ need to be continuous in the region V.)

Example 4   We'll recalculate the bee flux in Example 3 the easy way, using the divergence theorem. The theorem tells us that

$$\iint\limits_{S} \mathbf{F} \cdot \mathbf{n}\, dS = \iiint\limits_{V} \text{div } \mathbf{F}\, dV$$

so we can calculate the flux by integrating the divergence over the inside of the beehive cage. The divergence of the vector field $\mathbf{F}(x, y, z) = x\mathbf{i} + y\mathbf{j} + z\mathbf{k}$ is given by

$$\text{div } \mathbf{F} = \frac{\partial P}{\partial x} + \frac{\partial Q}{\partial y} + \frac{\partial R}{\partial z} = 1 + 1 + 1 = 3$$

So

$$\begin{aligned}
\text{Flux} &= \iiint\limits_{V} \text{div } \mathbf{F}\, dV \\
&= \iiint\limits_{V} 3\, dV \\
&= 3 \text{ volume } (V)
\end{aligned}$$

Since the volume of a ball of radius 1 is $(4/3)\pi$, this equals $(3)(4/3)\pi = 4\pi$. Just what we got before. Which way would you want to do the calculation if a swarm of bees was streaming toward you?

10.9   Stoking!

Stokes's theorem describes an amazing connection between line integrals and surface integrals of vector fields. It looks at a surface (called S) hanging around in space, which is not closed, but rather has an edge, or boundary, which is a curve (called C), as in Figure 10.9.

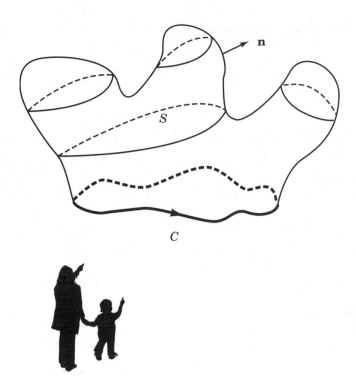

Figure 10.9 A surface S with boundary curve C and normal vector $\mathbf{n}$.

The unit normal vector to S is called $\mathbf{n}$. In its general form the theorem is:

Theorem (Stokes's Theorem)

$$\oint_C \mathbf{F} \cdot d r = \int\int_S (\text{curl } \mathbf{F} \cdot \mathbf{n})\, dS$$

Amazing! Adding up all the swirling of $\mathbf{F}$ along the surface S measures the amount by which the vector field is trying to twist the boundary curve of S. The total curl of $\mathbf{F}$ on the surface S is equal to the line integral of $\mathbf{F}$, which measures how much of the vector field is lined up along C. (Continuity is required of $\mathbf{F}$ and curl $\mathbf{F}$).

This theorem is really useful when one of these quantities is easier to calculate than the other. Stokes's theorem allows us to pick the easier path to the answer. Stokes's theorem requires the right choice of direction. If the vector field $\mathbf{n}$, which is perpendicular to the surface, points in the same direction as your head and you are walking along C, then the surface S has to be on your left. Mess this up and you get the answer with a minus sign in front of it.

There is a special case of Stokes's theorem when the surface S is a region in the plane. We'll talk first about this easy case. The unit normal vector $\mathbf{n}$ is then just $\mathbf{k}$, and the theorem becomes:

Theorem (Stokes's Theorem in the Plane) *Suppose D is a region in the plane with boundary curve C, oriented counterclockwise. Then*

$$\oint_C \mathbf{F} \cdot d\mathbf{r} = \iint_D (\text{curl } \mathbf{F}) \cdot \mathbf{k} \, dA$$

Example 1 (Escape from New York Ballet) A group of twirling ballet dancers has escaped opening night at the New York Ballet and is twirling through midtown Manhattan. The police, thinking quickly, throw a rope around the rotating dancers, holding them until their choreographers can come to claim them. As the spinning dancers come up against the rope, they exert a twisting force which pulls the rope around the group. The police know that they can maintain their control as long as the force along the rope is no more than 1000 (Newtons, if you care about units; Fig Newtons if you care about cookies). The force generated by the dancers is given by $\mathbf{F}(x, y) = (-2y)\mathbf{i} + (2x)\mathbf{j}$ (same units), and the rope is described by the circle C given by $x^2 + y^2 = 10$. Will the rope hold?

Solution The police realize that they need to calculate the line integral of the force $\mathbf{F}$ along the rope, and see whether this exceeds 1000. But they don't have time to parametrize the curve C and do the line integral calculations. Luckily, they know that they can make a simpler calculation using Stokes's theorem. The force pulling along the rope is equal to the total twirling going on inside:

$$\oint_C \mathbf{F} \cdot d\mathbf{r} = \iint_D (\text{curl } \mathbf{F}) \cdot \mathbf{k} \, dA$$

where C is the circle of radius 10 and D is the disk inside C.

Calculating $\nabla \times \mathbf{F}$ gives

$$\text{curl } \mathbf{F} = \nabla \times \mathbf{F} = \begin{vmatrix} \mathbf{i} & \mathbf{j} & \mathbf{k} \\ \dfrac{\partial}{\partial x} & \dfrac{\partial}{\partial y} & \dfrac{\partial}{\partial z} \\ -2y & 2x & 0 \end{vmatrix} = 4\mathbf{k}$$

and $(\text{curl } \mathbf{F}) \cdot \mathbf{k} = 4$. The function being integrated over D is just a constant equal to 4. Since the area of D is $\pi(10)^2 = 100\pi$, the total force pushing along the rope is 400π. This is greater than 1000 and the rope is not going to hold. Looks like a few whacks with the batons are going to be needed.

Well, that was an easy case of Stokes's theorem, since the surface S was flat on the xy-plane. You may notice that it looks suspiciously like Green's theorem. Stokes's theorem in the plane is pretty much the same as Green's theorem. But Stokes's theorem also works when S is a surface in space.

Example 2 (Postmodern Architecture)

The postpostmodernist architect Svenn Diagram has designed a roof shaped like the upside-down paraboloid $z = 16 - x^2 - y^2$ above the plane $z = 7$. The force of the wind on this roof is given by

$$\mathbf{F}(x, y, z) = 2yz\mathbf{i} + 2xz\mathbf{j} + 2xy\mathbf{k}$$

The roof is held down at its boundary C by bolts that will break if the integral $\oint_C \mathbf{F} \cdot d\mathbf{r}$ is greater than 6. Will the postpostmodernist movement blow away?

Solution We could directly calculate $\oint_C \mathbf{F} \cdot d\mathbf{r}$. But there's an easier way. Stokes's theorem tells us that

$$\oint_C \mathbf{F} \cdot d\mathbf{r} = \iint_D (\text{curl } \mathbf{F} \cdot \mathbf{n})\, dS$$

Let's calculate curl $\mathbf{F}$:

$$\nabla \times \mathbf{F} = \begin{bmatrix} \mathbf{i} & \mathbf{j} & \mathbf{k} \\ \dfrac{\partial}{\partial x} & \dfrac{\partial}{\partial y} & \dfrac{\partial}{\partial z} \\ 2yz & 2xz & 2xy \end{bmatrix} = \mathbf{0}$$

Now there's a good example of how Stokes's theorem can make things easy. The curl of $\mathbf{F}$ is zero, meaning (curl $\mathbf{F}$) $\cdot \mathbf{n}$ is zero, so integrating it over the roof will be pretty easy.

$$\oint_C \mathbf{F} \cdot d\mathbf{r} = \iint_S (\text{curl } \mathbf{F} \cdot \mathbf{n})\, dS = 0$$

The twisting forces of the wind combine to cancel out to zero, and the roof won't be torn off. Of course it still leaks, but that's a small price to pay for art.

Okay, that's it, the end of calculus. We're done. All that's left is the big end-of-calculus party, with the streamers, party hats and carbonated cider that is supposed to pass for champagne. Oh, yes, and I guess there's still the final. Oh, well, on to the next chapter . . .

What's Going to Be on the Final?

Here's where we try to help you figure out what will be on the final. While every course is different, there are certainly ways to make a good guess about what is going to be tested. The best source of information is the instructor. Pay close attention to what is said in class about what the exam covers. If you can get hold of exams from previous years, you will have a big advantage. It's a rare instructor who drastically changes the types of questions asked from one year to the next.

So here's a description of some of the common types of questions found on exams on the subjects discussed in this book.

If you want actual problems from real exams, with the answers carefully worked out, hey, we understand. A whole lot of these can be found at our Web site

www.howtoace.com

Please visit.

1. *Multiple integral problems* (most common; occur on 100% of final exams). These problems could involve computing total mass, averages, volumes,

or the center of mass. In general, they just tell you to compute the integral of some function over a region in the plane or in space. Often the hardest part of the problem is to describe the region correctly, using the right coordinate system. You will need to do two or three integrals to carry out a multiple integral calculation, and you will need to choose the right order for the variables that you integrate with respect to, depending on how you describe the region.

Trick Always check that your answer is reasonable. You can make very rough estimates to see if you are in the ballpark. If you are calculating the volume inside a ball of radius 1, then your answer should certainly be less than the volume of a cube enclosing that ball. Such a cube is $2 \times 2 \times 2$ and has volume 8, so if your answer is 214 you know you've got to redo the problem. If your volume calculation gives you a negative answer, that's also an indication to recheck your work. If your volume calculation gives you an answer like "$3x^2 - 7\pi y$," which has variables in it, then either you have not set up your integral correctly or you didn't integrate from the inside working out.

2. *Series convergence* (most common; occur on 100% of final exams). If your course covered sequences and series, you can be certain that there'll be a question, and probably a bunch of questions, that asks, "Does the following series converge?" You may or may not also be asked, "What does it converge to?"

   You should be on top of all the convergence tests, and on the difference between conditional and absolute convergence.

   Expect some tricky examples here, as most problems are easy to to do with the right test. Typically, some algebraic manipulations are required to get a series to fit into one of the tests.

3. *Partial derivatives and chain rule* (most common; occur on 90% of final exams). These problems ask you to compute partial derivatives of a function $f(x, y, z)$. A typical problem is:

   Compute $\partial f / \partial y$ for $f(x, y) = e^{xy} + y^x$.

   In a variation of this problem, you aren't told what the function is—you are just given some of the partial derivatives and asked to work out some others. You need to know the chain rule to do this.

4. *Sketch and parametrize regions in the plane or solids in space* (very common; occur on 80% of final exams). This kind of problem involves describing with formulas something like "the region between a cylinder of radius 1 with axis the z-axis and a sphere of radius 2 centered at the origin." They are often associated with multiple integral problems over

these regions. Using the right kind of coordinates, cylindrical, Cartesian or spherical, is the key to setting these up.

5. *Computations with vectors* (very common; occur on 80% of final exams). These problems could ask you to compute dot products, cross products, projections of one vector on another, areas of triangles and/or parallelograms, equations of lines or planes. A basic understanding of the geometric meaning of vectors together with a few formulas like $\mathbf{a} \cdot \mathbf{b} = |\mathbf{a}||\mathbf{b}| \cos \theta$ should get you through without too much trouble.

6. *Graph a surface* (very common; occur on 80% of final exams). You may be asked to graph one of the quadric surfaces like a paraboloid, ellipsoid, hyperboloid, or hyperbolic paraboloid. Or you may be asked to graph some kind of cylinder generated when one of the variables does not appear in the equation. Or you may be given some random function $z = f(x, y)$ and have to graph that.

7. *State theorems and definitions* (very common; occur on 80% of final exams). Although few instructors expect you to prove theorems on exams, they often expect you to be able to state theorems and definitions. Typical questions might ask for the statement of Green's theorem, or Stokes's theorem, or the integral test for convergence of series. Or you might be asked for the definition of a partial derivative, or precisely what it means to say that a series converges, or to say that $f(x, y)$ is continuous at (x_0, y_0). For full credit, you need to be sure to state the theorem with the right amount of precision. So a statement of Green's theorem should include the continuity assumptions if they were discussed in your course.

8. *Line integrals* (very common; occur on 80% of final exams). These problems describe a curve and a function or vector field. You need to set up and compute a line integral. Sometimes you will need to compute directly, and sometimes this is hopeless and you need to use Green's theorem to do the computation.

9. *Find critical points* (very common; occur on 80% of final exams). Given a function $f(x, y)$ or $f(x, y, z)$, you will need to compute its critical points and determine their nature: max, min, or saddle.

10. *Solve a max/min problem* (very common; occur on 80% of final exams). The goal is to find the absolute maximum or minimum of some function with a constraint. You can either use the constraint to lower the number of variables by 1, or use Lagrange multipliers. To get full credit, use some kind of argument to show that the point you have found really is a max or min.

11. *Compute Taylor series* (pretty common; occur on 70% of final exams). You're given a function and asked to compute its Taylor series expansion. Sometimes you are also asked where the series converges.

Variation 1: Compute the Taylor series with powers of $(x - a)$ instead of just powers of x.

Variation 2: Compute a few terms and then work out the formula for the error, or remainder, if you stop there.

12. *Compute the radius of convergence of a power series* (pretty common; occur on 70% of final exams). You're given a power series:

$$\sum_{n=0}^{\infty} a_n(x - a)^n = a_0 + a_1(x - a) + a_2(x - a)^2 + \cdots + a_n(x - a)^n + \cdots$$

and need to work out the radius r of the interval $(a - r, a + r)$ around a in which it converges. Could be as little as 0 or as big as ∞. Often, you are asked to check convergence at the endpoints of the interval as well. This is automatically implied when you are asked for the interval of convergence.

13. *Computing with vector fields* (pretty common; occur on 70% of final exams). You need to know the definitions of div and curl and be able to determine whether a vector field is conservative. Answering these questions entails knowing what the words mean and how to compute.

14. *Surface integrals and flux* (pretty common; occur on 70% of final exams). Some sort of surface is described, often part of a plane, cylinder, or sphere. You're told to find the flux of some vector field across this surface.

    Be sure to use the coordinate system that's best fitted to the surface. Often there is a big shortcut for these problems if you can use the divergence theorem or Stokes's theorem.

15. *Set up and compute volumes of three-dimensional regions* (pretty common; occur on 70% of final exams). Setting up the integral is often the tricky part. Sometimes the regions are bounded by the graphs of functions, and sometimes the region is described in words, for example, "The solid which is inside the cylinder of radius 2 with axis the z-axis, above the plane $z = 0$ and below the plane $x + z = 4$." Sketching the region is a good idea.

    Calculating volumes can be either a job for a triple integral or for a double integral. If you are using a triple integral, use spherical or cylindrical coordinates if the shape is easier to describe in these.

16. *Problems on limits and continuity* (pretty common; occur on 70% of final exams). You saw these problems for one-variable functions $f(x)$. Now you see them asked about functions $f(x, y)$ of two variables and even functions of three variables $f(x, y, z)$. Also asked are problems about where $f(x, y)$ or $f(x, y, z)$ is continuous.

17. *Directional derivatives and the gradient* (pretty common; occur on 70% of final exams). You might expect to have to compute a particular directional derivative $D_{\mathbf{u}}f$. Be sure that the vector $\mathbf{u}$ is a unit vector. Or you may have to find the direction of steepest ascent or descent.

18. *Green's theorem, the divergence theorem, and Stokes's theorem* (common; occur on 50% of final exams). In these problems you are explicitly instructed to calculate an integral using a specific theorem. This is actually easier than having to figure out for yourself how to do it. But you will need to bone up on these theorems, which means work through some examples in each of them, and know what they say really well.

Sometimes you are asked to calculate an integral two different ways to verify Green's theorem. So you need to understand how to calculate the expressions on both sides of the equation.

Whenever you are asked to compute the flux through a surface, you should always ask yourself if these theorems can help you.

19. *Velocity and acceleration vectors for parametric curves* (common; occur on 50% of final exams). You may be given any of the position, velocity, and acceleration vectors, plus some initial information, and be expected to figure out the other two. Lots of potential for projectile problems here.

20. *Center of mass, and moments* (common; occur on 50% of final exams). These problems involve performing multiple integrals to calculate centers of mass, moments of inertia, average density, and so on. You need to know what the words mean, what the formulas are, and how to set up and calculate the multiple integrals. For example, if you are asked to calculate the moment of inertia around the z-axis of a region V in space whose density is given by $\sigma(x, y, z)$, you need to know that the formula is

$$\int\!\!\int_V\!\!\int \sigma(x, y, z)\,(x^2 + y^2)\,dx\,dy\,dz$$

Of course, you also need to know how to set up and calculate this multiple integral.

21. *Improper integrals* (common; occur on 50% of final exams). Integrals where one or both of the integration limits are infinite. Or where the function has an infinitely high spike. Handle these with an appropriate limit.

22. *Lagrange multipliers* (common; occur on 50% of final exams). Not covered in all courses, but if you've seen it in class it's very likely to be on the final. Lagrange multipliers are among the most common applications of calculus in areas like economics, so if you're in a business calculus course the likelihood of such a question shoots up.

23. *L'Hôpital's rule* (common; occur on 50% of final exams). Limits of functions which give indeterminate forms (like $0/0$ or ∞/∞) when you plug in. Use L'Hôpital's rule, but be careful that the rule really applies. Quite commonly you have to use the rule two or three times. Be careful not to use the rule if the ratio is not indeterminate.

24. *Tangent planes or tangent lines* (not too common; occur on 25% of final exams). A typical problem asks you to find the equation of the tangent plane to a surface at a given point. You need to compute the partial derivatives, use them to find vectors in the directions of the corresponding tangent lines, and take their cross product to get a normal vector. The equation of the plane can then be found by taking all vectors perpendicular to the normal vector.

25. *Arc length* (not too common; occur on 25% of final exams). This isn't so common, because there are very few examples that can be done by hand. The integrals usually get too ugly. But a simple example of arc length along a line, circle, or helix could occur.

26. *Curvature* (uncommon; occur on 10% of final exams). Again, here, the examples get so nasty so quickly, these problems don't make for good exam questions.

GLOSSARY: A Quick Guide to the Mathematical Jargon

Absolute Convergence We tried to get that vodka company to put an ad here but they wouldn't go for it. They don't understand the vodka-drinking potential of our readers. Anyway, a series $\sum a_n$ converges absolutely if its absolute value $\sum |a_n|$ converges. This is at least as difficult to achieve as convergence of the original series, so absolute convergence implies convergence. A series which is convergent but not absolutely convergent is called *conditionally convergent*.

Acceleration Vector The derivative of the velocity vector or the second derivative of the position vector. Its length tells you how fast speed is changing. Useful when shmoozing another driver at a red light. "I was so impressed with your acceleration vector back there."

Arc Length Reputed by Noah to have come in at 300 cubits. To calculate the length of a parametrized curve, we compute a line integral. The quantity $d s$ that we integrate to get the length is called the element of arc length. For a parametrized curve $\mathbf{C}(t) = \langle x(t), y(t), z(t) \rangle$, $a \leq t \leq b$, the element of arc length is $d s = \sqrt{x'(t)^2 + y'(t)^2 + z'(t)^2}\, d t$ and the length of the curve is
$$\int_C d s = \int_a^b \sqrt{x'(t)^2 + y'(t)^2 + z'(t)^2}\, d t.$$

Cardioid Derogatory name for a professional greeting card designer. Also a heart-shaped graph which in polar coordinates is given by equations like $r = 1 + \cos \theta$ or $r = 1 + \sin \theta$.

Cardioid Arrest Happens to the hearts of those who open up the final and haven't read this book.

Center of Mass That spot right in the center of your belly after a massive Thanksgiving dinner. As you move, it moves, too, but more slowly. For a two-dimensional flat object, the point on which it can be balanced. For a three-dimensional solid object, if you slice the object along any plane through this point, the two halves have the same mass.

Centroid Fancy name for the center of mass. Also a unit of currency in Detroid.

Chain Rule (for Functions of Several Variables) A rule known to teenagers who look like they've been attacked by a staple gun. Also a way of taking the derivative of composite functions such as $f(g(x, y), h(x, y))$, which are themselves functions of more than one variable.

Conditional Convergence   The kind of joining in holy matrimony where one person proposes by saying, "If you want to marry rich old me, you must sign a prenup." Also, a series with some terms possibly negative, which converges. The key example is the series $1 - \frac{1}{2} + \frac{1}{3} - \frac{1}{4} + \cdots$ which converges conditionally but not absolutely.

Continuity (for Functions of Several Variables)   Just as with one variable, a function $f(x, y)$ is continuous at (x_0, y_0) if

$$\lim_{(x, y) \to (x_0, y_0)} f(x, y) = f(x_0, y_0)$$

and continuous if it is continuous for each (x_0, y_0) in its domain.

Contour Curve   Also known as the contour de France curve, it's the path of a famous bicycle race held in France by convicted felons. (Con Tour, get it? Yikes.) Also a "level curve" of a function, a curve where the values of the function are constant. For a function $z = f(x, y)$, the (x, y) values where $f(x, y) - 3$ describe the $z - 3$ contour curve.

Convergence (of a Sequence)   Are the terms in the sequence zeroing in on some number, or in other words, do they have a limit? If yes, we say that the sequence converges to that number.

Convergence (of a Series)   When all the planets and stars of a series line up in the heavens, and its number comes up. Made precise with partial sums. Set S_n to be the sum of the first n terms of the series, and check whether the sequence S_n converges.

Convergence (of an Improper Integral)   Sometimes an ill-brought-up integral, badly mannered and not adhering to the usual rules of integral society, can be tamed and civilized. Such integrals can have upper and lower limits which are $\pm\infty$, rather than the more civilized a and b, or have functions in them like $1/\sqrt{x}$, which is not defined at $x = 0$. Still, with appropriate handling they can be interpreted as limits of standard type integrals. Thus $\int_0^1 \frac{1}{\sqrt{x}} \, dx$, which has a nasty singularity at $x = 0$, is interpreted as the standard $\lim_{\epsilon \to 0} \int_\epsilon^1 \frac{1}{\sqrt{x}} \, dx$. This can be easily evaluated:

$$\lim_{\epsilon \to 0} \int_\epsilon^1 \frac{1}{\sqrt{x}} \, dx = \lim_{\epsilon \to 0} \int_\epsilon^1 x^{-1/2} \, dx = \lim_{\epsilon \to 0} [2\sqrt{x}]_\epsilon^1 = \lim_{\epsilon \to 0} [2 - 2\sqrt{\epsilon}] = 2$$

Cross Product   A very grumpy product, probably because its formula is so much more complicated than the dot product's. But the cross product of two vectors is a vector, not a number, so it has to have a lot of coordinates. When $\mathbf{v} = \langle v_1, v_2, v_3 \rangle$ and $\mathbf{w} = \langle w_1, w_2, w_3 \rangle$, then $\mathbf{v} \times \mathbf{w} = (v_2 w_3 - v_3 w_2)\mathbf{i} + (v_3 w_1 - v_1 w_3)\mathbf{j} + (v_1 w_2 - v_2 w_1)\mathbf{k}$. Remember this using a determinant formula:

$$\mathbf{v} \times \mathbf{w} = \det \begin{bmatrix} \mathbf{i} & \mathbf{j} & \mathbf{k} \\ v_1 & v_2 & v_3 \\ w_1 & w_2 & w_3 \end{bmatrix}$$

Curl   What happens to your hair when you see a Stokes's theorem problem on the final. But no need to panic, just remember that curl measures the tendency of a vector field to spin around an axis. For a vector field $\mathbf{V}$ in 3-space, curl $\mathbf{V}$ is a vector field, too. At a particular point in space, the vector field curl $\mathbf{V}$ lines up along the axis where the most spinning around is taking place and its length gives the amount of spinning. More spinning, bigger curl. The direction that curl $\mathbf{V}$ points along the axis of maximal rotation is given by the right-hand rule. Well, with all that, there's a simple formula to calculate curl $\mathbf{V}$:

$$\text{curl } \mathbf{V} = \nabla \times \mathbf{V}$$

Curvature   Derived from "curvy chair," a chair style out of the nineteenth century's premodern school. Winner of many design awards, its adoption was hampered by its tendency to flip upside down when someone sat in it. Also a measure $\kappa(t)$ of the bending of a parametrized curve $\mathbf{r}(t)$. Curvature tells us how quickly the unit tangent vector to the curve,

$$T(t) = \frac{\mathbf{r}'(t)}{|\mathbf{r}'(t)|}$$

is changing direction. For a unit speed curve $\mathbf{r}(s)$ the unit tangent vector is just $\mathbf{r}'(s)$ and the formula for $\kappa(s)$ is

$$\boxed{\kappa(s) = |\mathbf{r}''(s)|}$$

The curvature of an arbitrary curve $\mathbf{r}(t)$ is given by

$$\boxed{\kappa(t) = \frac{|\mathbf{T}'(t)|}{|\mathbf{r}'(t)|}}$$

Cylindrical Coordinates   A hybrid. Just take polar coordinates r and θ in the plane and tack on a z from rectangular coordinates. Looks like a Volkswagon with the hood from a Cadillac.

Del   The symbol ∇ is called del, and stands for

$$\nabla = \frac{\partial}{\partial x}\mathbf{i} + \frac{\partial}{\partial y}\mathbf{j} + \frac{\partial}{\partial z}\mathbf{k}$$

∇ is an operator—by itself ∇ doesn't really mean anything, but ∇f is the gradient of a function $f(x, y, z)$, $\nabla \cdot \mathbf{F}$ is the divergence of the vector field $\mathbf{F}$ and $\nabla \times \mathbf{F}$ is the curl of the vector field $\mathbf{F}$. Just as you can't get a good sandwich until you add an "i" to "del" to get deli, so you have to add a function or vector field to Del to get a meaningful result.

Determinant   A number associated to a square array of numbers. Very useful in calculating the cross product of two vectors.

$$\det \begin{bmatrix} a & b \\ c & d \end{bmatrix} = \begin{vmatrix} a & b \\ c & d \end{vmatrix} = ad - bc$$

$$\det \begin{bmatrix} a & b & c \\ d & e & f \\ g & h & j \end{bmatrix} = a\begin{vmatrix} e & f \\ h & j \end{vmatrix} - b\begin{vmatrix} d & f \\ g & j \end{vmatrix} + c\begin{vmatrix} d & e \\ g & h \end{vmatrix}$$

Differential Equation   An equation that contains derivatives, such as $3\dfrac{\partial^2 x}{\partial^2 t} - 5\dfrac{\partial x}{\partial t} - x = 7$. Despite their names, these equations are often very rude, not deferential at all.

Direction Angles   The best angles at which to get a closeup when shooting a calculus movie. Contact authors if interested in this book's movie rights. Also, if we pick an arbitrary direction in space, the angles α, β, and γ between that direction and the x-, y-, and z-axes are called the direction angles.

Direction of Steepest Ascent/Descent   The compass direction in which you should head in order to go up or down a hill the steepest way possible. The vector ∇f points in the direction of steepest ascent and $-\nabla f$ points in the direction of steepest descent. So you should not head in the direction of $-\nabla f$ when standing on the edge of a cliff.

Directional Derivative   Men never ask for these. The directional derivative of a function $f(x, y)$ in the direction of a unit vector $\mathbf{u}$ is by

$$D_{\mathbf{u}}f = \nabla f \cdot \mathbf{u}$$

Divergence, or div   Two paths diverged in a wood, and I took the one less traveled by, and got the wrong answer. The divergence of a vector field $\mathbf{F}$ at a point is a number which measures the amount by which the vector field is compressing or expanding at that point. The formula is

$$\text{div } \mathbf{F} = \nabla \cdot \mathbf{F}$$

Divergence Theorem The divergence theorem in the plane states that if you add up (by integrating) the divergence of a vector field in a region in the plane (measuring its overall total expansion or compression in the region), you get the total flow of the vector field across the boundary, called the flux. Extremely useful if you are in the process of spilling large quantities of maple syrup onto a raised stage with limited drainage.

Theorem:
$$\iint_D \text{div } \mathbf{F} \, dx \, dy = \oint_C \mathbf{F} \cdot \mathbf{n} \, ds$$

The version for 3-space says that the total divergence of the vector field in a region in 3-space equals the flow across the surface which is the boundary of the region.

Theorem:
$$\iiint_V \text{div } \mathbf{F} \, dx \, dy \, dz = \iint_S \mathbf{F} \cdot \mathbf{n} \, dS$$

Continuity is needed for $\mathbf{F}$ and div $\mathbf{F}$.

Dot Product Early internet startup, crashed and burned. Also a way of getting a number by taking a product of two vectors which measures how much they line up; it is the product of their lengths times the cosine of the angle between them. Very useful because of its supereasy formula, which makes it a cinch to calculate. When $\mathbf{v} = \langle v_1, v_2, v_3 \rangle$ and $\mathbf{w} = \langle w_1, w_2, w_3 \rangle$, then

$$\mathbf{v} \cdot \mathbf{w} = v_1 w_1 + v_2 w_2 + v_3 w_3$$

It works just the same for vectors with more or less components.

Double Integral Served with ice, it's a favorite cocktail drink at math parties, right up there with the Pink Lagrange Multiplier. Also an iterated integral, performed twice. When you see

$$\int_a^b \int_c^d f(x, y) \, dx \, dy$$

you should think of doing two ordinary integrals, starting with the inside one $\int_c^d f(x, y) \, dx$, and continuing by plugging the answer from that one into the outside limits. The inner limits can be functions but the outer limits are almost always constants.

Ellipsoid   Object removed during elliposuction. Also the surface which looks like a deformed sphere, and whose equations are

$$\frac{x^2}{a^2} + \frac{y^2}{b^2} + \frac{z^2}{c^2} = 1$$

Flux   State of your calculus knowledge—usually increasing, sometimes decreasing, but in a state of change. Also a measure of the amount of a vector field that is flowing through a surface. At each point in the surface, the flux is given by $\mathbf{V} \cdot \mathbf{n}$, where $\mathbf{V}$ is a vector field and $\mathbf{n}$ is the unit vector perpendicular to the surface at that point. Actually, there are two unit vectors, and we must choose a direction perpendicular to the surface, or an orientation, to calculate the flux. The total flux of a vector field through a surface is related to the divergence of the vector field integrated over the region inside the surface by the divergence theorem.

Geometric Series   A series of the form $a + ar + ar^2 + ar^3 + \cdots$. Converges to $\dfrac{a}{1-r}$ if $|r| < 1$ and diverges otherwise. Great for modeling distance traveled by bouncing balls.

Gradient, or grad   Good with turkey and mashed potatients. The gradient of a function $f(x, y, z)$ is a vector in space:

$$\nabla f(x, y, z) = \frac{\partial f}{\partial x}\mathbf{i} + \frac{\partial f}{\partial y}\mathbf{j} + \frac{\partial f}{\partial z}\mathbf{k}$$

Green's Theorem   *Version 1:* Most environmental activists drive SUVs, have redwood decks, and use disposable diapers, even if they don't have kids.

Version 2: In the plane, Green's theorem tells us a way to calculate the line integral around a closed curve, when we go around it counterclockwise. If $\mathbf{F} = P\mathbf{i} + Q\mathbf{j}$ is a vector field and C goes around a region R, then

$$\oint_C P\,dx + Q\,dy = \iint_R \left(\frac{\partial Q}{\partial x} - \frac{\partial P}{\partial y}\right) dx\,dy$$

Continuity is needed for P, Q, and their derivatives.

Harmonic Series $1 + \frac{1}{2} + \frac{1}{3} + \cdots$. One of the great series of our times. Although the terms get smaller and smaller, the series does NOT converge.

Hyperboloid   Very active boloid, can't keep still. Also the surface which looks like a hyperbola spun around an axis, and whose equations are like the typical example:

$$x^2 + y^2 - z^2 = 1$$

Indeterminate Form   What you use to file your income taxes. Also a limit which approaches $0/0$ or ∞/∞. Great opportunity to use L'Hôpital's rule.

Infinite Sequence   Not to be confused with infinite sequins, garments worn by pianists in Las Vegas. An infinite sequence is an unending list of numbers. The most famous is $1, 2, 3, 4, \ldots$ but $2, 4, 6, 8, \ldots$ and $1, 4, 9, 16, \ldots$ are good examples, too.

Lagrange Multipliers   A pocket tool which performs 45 functions. Also a technique for solving constrained max-min problems that's useful in a variety of disciplines, particularly economics where investments are constrained by the total capital available, unless you're Bill Gates. When maximizing f holding g constant, find the points where ∇f is a constant multiple of ∇g. This will give the points where f is maximized or minimized subject to the constraint.

Line Integral   An integral of a function defined on a parametrized curve in space or in the plane.

Maclaurin Series   Power series which looks like

$$f(0) + f'(0)x + \frac{f''(0)}{2!}\, x^2 + \frac{f'''(0)}{3!}\, x^3 + \cdots$$

Can be used to approximate a function $f(x)$ by a polynomial near $x = 0$.

Moment of Inertia   That split second that occurs right after the words "Let's get married." Also a measure of the resistance to changing the angular momentum of an object spinning around an axis. The formula for the moment of inertia of a region R in the plane with density $\sigma(x, y)$ around the x-axis is

$$I_x = \iint\limits_{R} y^2 \sigma(x, y)\, dA$$

and around the y-axis it's

$$I_y = \iint\limits_{R} x^2 \sigma(x, y)\, dA$$

The moment of inertia around the z-axis of a region V whose density is given by $\sigma(x, y, z)$ is

$$\iiint\limits_V (x^2 + y^2)\, \sigma(x, y, z)\, dx\, dy\, dz$$

Normal Vector Just your everyday, unassuming, perpendicular vector. For a curve in the plane a normal vector is perpendicular to the tangent line of the curve; for a surface in 3-space it's perpendicular to the tangent plane. Occasionally goes by the name orthogonal vector.

Paraboloid A couple of boloids. Also the surface which looks like a parabola spun around an axis. The generic example is $z = x^2 + y^2$.

Paranoiboloid A boloid that thinks it's under microwave surveillance by Martians.

Parametric Curve A couple of metric curves. Also a vector-valued function that describes the position of a moving particle/bug/baseball/wedge of Gouda cheese in space. Typically looks like

$$\mathbf{C}(t) = \langle x(t), y(t), z(t) \rangle \qquad a \le t \le b$$

Partial Derivative An incomplete answer to a derivative question on an exam. May cost you points. But may get you partial credit. Also a derivative of a function of more than one variable by one of the variables. Written like $\partial f(x, y)/\partial x$ or $\partial f(x, y, z)/\partial y$.

Polar Coordinates A fashion style popular in the far north, features matching white on white pant suits. Also a way of describing the location of points in the plane by (r, θ), their distance from a point and their angle from the positive x-axis. The angle θ from the axis is measured counterclockwise.

Polar Bear A really really difficult polar coordinates problem.

Position Vector Important chapter in the vector sex manual. Also a vector that describes the location of an object.

Potential Function An on-campus party for which a permit must still be obtained. Also a function f, the gradient of which gives a vector field $\mathbf{F}$.

Power Series Baseball championship with double digit scores. Also an "infinite order polynomial," which looks like

$$a_0 + a_1x + a_2x^2 + a_3x^3 + \cdots$$

or more generally,

$$a_0 + a_1(x - a) + a_2(x - a)^2 + a_3(x - a)^3 + \cdots$$

Radius of Convergence A power series

$$a_0 + a_1(x - a) + a_2(x - a)^2 + a_3(x - a)^3 + \cdots$$

will converge if $|x - a|$ is less than this radius, and diverge if $|x - a|$ is larger. Can be anything from 0 to ∞. At points where $|x - a|$ is equal to the radius of convergence, the power series can either diverge or converge, and the points must be tested separately.

Saddle Also known as a hyperbolic paraboloid. As in, "Time to ride, pardner. Throw a hyperbolic paraboloid on that horse and let's hit the trail." It's a surface that bends in two different directions, like a, well you know. A typical example is $z = x^2 - y^2$.

Scalar Fancy word for number, used by snobbish numbers to emphasize that they are NOT vectors.

Spherical Coordinates Alternate coordinates for 3-space where ρ is the length of the line segment from you back to the origin, ϕ is the angle of that line segment with the positive z-axis, and θ is the angle between the projection of the segment to the xy-plane and the positive x-axis. People who believe in the mystical power of pyramid-shaped crystals tend to give their coordinates in spherical, often with a reincarnated spirit named Ocnomon at the origin.

Stokes's Theorem A theorem celebrated each October with Stokestoberfest, wherein lots of Bavarians drink gallons of beer in a roped-off area called a beergarden D. As they drink, they sing and dance, with a lot of spinning, which puts tremendous strain on the rope C. The theorem says:

$$\oint_C \mathbf{F} \cdot d\mathbf{r} = \iint_D (\text{curl } \mathbf{F}) \cdot \mathbf{k} \, dA$$

There's also a version for 3-space. Different Stokes for different folks. Here we look at a surface S hanging around in space whose edge, or boundary, is a curve called C. Take $\mathbf{F}$ to be a vector field on S and call the unit normal vector $\mathbf{n}$. The theorem equates a line integral over the curve C and a surface integral over S:

$$\oint_C \mathbf{F} \cdot d\mathbf{r} = \iint_S \text{curl } \mathbf{F} \cdot \mathbf{n} \, dS$$

The total twisting effect of the vector field over the surface S is equal to the total tangential part of the vector field along the curve C. In both versions continuity is needed for $\mathbf{F}$ and curl $\mathbf{F}$.

Surface Integral   Some integrals have complex personalities, but to the world at large they show a superficial persona, never hinting at the deep struggles which consume them. These integrals allow us to calculate quantities associated to surfaces in space. To calculate an integral of a function on a surface, we use a double integral involving the element of surface area dS.

Tangent Line to a Parametrized Curve   Where tangents go to get their cars registered, buy stamps, or check in their luggage. Also a line which approximates a curve $\mathbf{r}(t) = \langle x(t), y(t), z(t) \rangle$ near a point $\mathbf{r}(t_0)$. Given by the parametric equation

$$\langle x, y, z \rangle = \langle x_0, y_0, z_0 \rangle + s\mathbf{r}'(t_0) \qquad -\infty < s < \infty$$

or by the equation

$$\frac{x - x_0}{x'(0)} = \frac{y - y_0}{y'(0)} = \frac{z - z_0}{z'(0)}$$

Tangent Plane   A plane in space that is parallel to a surface $\mathbf{r}(u, v)$ in space at a given point $\langle x_0, y_0, z_0 \rangle = \mathbf{r}(u_0, v_0)$, approximating the surface near that point as best as any plane can. Use the cross product to get a normal vector $\mathbf{n} = \mathbf{r}_u \times \mathbf{r}_v$ at the point (x_0, y_0, z_0). Then the equation of the plane is

$$[(x - x_0)\mathbf{i} + (y - y_0)\mathbf{j} + (z - z_0)\mathbf{k}] \cdot \mathbf{n} = 0$$

Tangenital Component   A particularly unfortunate misspelling of tangential component, which is that part of a vector that is in the direction of the tangent to a curve. Often applied to the acceleration vector, where the tangential component measures how fast the speed is changing.

Taylor Series   The finals in the alteration competition at the Tailor's Olympics. Also the infinite series associated to a function $f(x)$ and a point $x = a$.

$$f(a) + f'(a)(x - a) + \frac{f''(a)}{2!}(x - a)^2 + \frac{f'''(a)}{3!}(x - a)^3 + \cdots$$
$$+ \frac{f^{(n)}(a)}{n!}(x - a)^n + \cdots$$

A finite number of terms give a way to approximate $f(x)$. The error can be bounded using Taylor's remainder formula.

Triple Integral   As triple play is to double play, so triple integral is to double integral. You do the same sort of thing but one more time. You calculate these by iterating the integral process three times, starting with the inside and working out.

Triple Scalar Product   Ice cream treat sometimes served in the calculus lounge. Also a way of multiplying three vectors **u**, **v**, **w** to get a number which is the volume of the parallelepiped they span. Calculated by taking $(\mathbf{u} \times \mathbf{v}) \cdot \mathbf{w}$.

Velocity Vector   A vector which gives the speed and direction of a particle moving along a parametrized path $\mathbf{r}(t) = \langle x(t), y(t), z(t) \rangle$. The formula is simple:

$$\mathbf{v}(t) = \mathbf{r}'(t) = \langle x'(t), y'(t), z'(t) \rangle$$

Zeno   The last entry in any dictionary of calculus terms. Zeno was a Greek philosopher best known for Zeno's paradox. He pointed out that for a runner to go a mile, first she must go half the distance, then half the remaining distance, then half the remaining distance ad infinitum. Since clearly the runner cannot perform an infinite number of steps in a finite amount of time, motion is an impossiblility and is therefore an illusion. We ended our last book by saying that therefore all the world is just a dream and you should roll over and go back to sleep. However, in this book we have seen that an infinite number of distances can add up to a finite quantity. In fact, $1/2 + 1/4 + 1/8 + \ldots = 1$ mile. So much for Zeno's paradox. Wake up and smell the coffee.

INDEX

How to Ace the Rest of Calculus: Just the Facts

L'Hôpital's Rule:

If $\lim\limits_{x \to a} \dfrac{f(x)}{g(x)}$ is an indeterminate form $\left(\dfrac{0}{0} \text{ or } \dfrac{\infty}{\infty}\right)$, then

$$\lim_{x \to a} \frac{f(x)}{g(x)} = \lim_{x \to a} \frac{f'(x)}{g'(x)}$$

Improper Integrals:

* $\displaystyle\int_a^\infty f(x)\,dx = \lim_{b \to \infty} \int_a^b f(x)\,dx$

* $\displaystyle\int_{-\infty}^b f(x)\,dx = \lim_{a \to -\infty} \int_a^b f(x)\,dx$

* If $f(x)$ is undefined at a, we let

$$\int_a^c f(x)\,dx = \lim_{b \to a^+} \int_b^c f(x)\,dx$$

Polar Coordinates:

* $x = r\cos\theta$
  $y = r\sin\theta$
  $r = \sqrt{x^2 + y^2}$

  Also, $\tan\theta = y/x$, assuming $x \neq 0$.

* The equation of a circle of radius a centered at $(0, a)$ is $r = 2a\sin\theta$.

* The equation of a circle of radius a centered at $(a, 0)$ is $r = 2a\cos\theta$.

* Cardioids are given by $r = a(1 \pm \sin\theta)$ or $r = a(1 \pm \cos\theta)$.

* Area A bounded by two radial lines given by $\theta = \alpha$ and $\theta = \beta$, and by the curve $r = f(\theta)$, for $\alpha \le \theta \le \beta$ is

$$A = \int_\alpha^\beta \tfrac{1}{2}[f(\theta)]^2\,d\theta$$

* Area A bounded by the two radial lines $\theta = \alpha$ and $\theta = \beta$ and the two curves $r = f(\theta)$ and $r = g(\theta)$, where $0 \leq f(\theta) \leq g(\theta)$ for all θ between α and β, is $A = \dfrac{1}{2} \displaystyle\int_{\alpha}^{\beta} [g(\theta)]^2 - [f(\theta)]^2 \, d\theta$.

Infinite Series:

A series $\displaystyle\sum_{n=1}^{\infty} a_n = a_1 + a_2 + a_3 + \cdots$ *converges to* S if $\lim\limits_{n \to \infty} S_n = S$, where $S_n = a_1 + a_2 + \cdots + a_n$ is the *nth partial sum*. In this case, we write $\displaystyle\sum_{n=1}^{\infty} a_n = S$.

Geometric Series:

The *geometric series* $a + ar + ar^2 + ar^3 + \cdots$ converges to $\dfrac{a}{1-r}$ if $|r| < 1$ and diverges if $|r| \geq 1$.

p-Series:

A *p-series* $\displaystyle\sum_{n=1}^{\infty} \dfrac{1}{n^p}$ converges for $p > 1$ and diverges for $p \leq 1$.

Absolute Convergence:

* A series $\displaystyle\sum_{n=1}^{\infty} a_n$ is *absolutely convergent* if the series $\displaystyle\sum_{n=1}^{\infty} |a_n|$ converges.

* If a series is absolutely convergent, it is convergent.

* If a series converges but is not absolutely convergent, we say it is *conditionally convergent*.

Tests for Convergence:

* nth-Term Test: If $\lim\limits_{n \to \infty} a_n \neq 0$, then the infinite series $\sum\limits_{n=1}^{\infty} a_n$ diverges.

* Integral Test: If $a_n = f(n)$, where f is positive-valued, continuous, and decreasing for $x \geq 1$, then $\int_1^{\infty} f(x)\,dx$ and $\sum\limits_{n=1}^{\infty} a_n$ are partners in crime; either both converge or both diverge.

* Basic Comparison Test: If $\sum a_n$ and $\sum b_n$ are positive term series, and if $b_n \geq a_n$ for all n, then:

  1. If $\sum b_n$ converges, then so does $\sum a_n$.

  2. If $\sum a_n$ diverges, then so does $\sum b_n$.

* Limit Comparison Test: If $\sum a_n$ and $\sum b_n$ are positive term series, and for some positive number k,

$$\lim_{n \to \infty} \frac{a_n}{b_n} = k$$

  then either both series converge or both series diverge.

* Alternating Series Test: If $a_1 - a_2 + a_3 - a_4 + \cdots + (-1)^{n+1}a_n + \cdots$ or $-a_1 + a_2 - a_3 + a_4 + \cdots + (-1)^n a_n + \cdots$ are alternating series, where each $a_n > 0$, $a_{n+1} \leq a_n$ for all n, and $\lim\limits_{n \to \infty} a_n = 0$, then the alternating series converges.

* Ratio Test: If $\sum\limits_{n=1}^{\infty} u_n$ is a series, then

  1. If $\lim\limits_{n \to \infty} \left| \dfrac{u_{n+1}}{u_n} \right| < 1$, then the series converges absolutely.

  2. If $\lim\limits_{n \to \infty} \left| \dfrac{u_{n+1}}{u_n} \right| > 1$, then the series diverges.

  3. If $\lim\limits_{n \to \infty} \left| \dfrac{u_{n+1}}{u_n} \right| = 1$, then we don't know anything.

* Root Test: If $\displaystyle\sum_{n=1}^{\infty} u_n$ is a series, then:

  1. If $\displaystyle\lim_{n \to \infty}(|u_n|)^{1/n} < 1$, then the series converges absolutely.

  2. If $\displaystyle\lim_{n \to \infty}(|u_n|)^{1/n} > 1$, then the series diverges.

  3. If $\displaystyle\lim_{n \to \infty}(|u_n|)^{1/n} = 1$, then we don't know anything.

Power Series:

A *power series* is a series of the form

$$\sum_{n=0}^{\infty} a_n x^n = a_0 + a_1 x + a_2 x^2 + \cdots + a_n x^n + \cdots$$

Use the ratio test to determine *radius of convergence* and check endpoints to determine the *interval of convergence*, which is the set of all values of x where it converges.

Taylor Series:

* Taylor's Approximation:

$$f(x) \approx f(a) + f'(a)(x - a) + \frac{f''(a)}{2!}(x - a)^2 + \frac{f'''(a)}{3!}(x - a)^3$$
$$+ \cdots + \frac{f^{(n)}(a)}{n!}(x - a)^n$$

* Maclaurin's Approximation:

$$f(x) \approx f(0) + f'(0)x + \frac{f''(0)}{2!}x^2 + \frac{f'''(0)}{3!}x^3 + \cdots + \frac{f^{(n)}(0)}{n!}x^n$$

* Taylor's Formula with Remainder:

$$f(x) = \underbrace{f(a) + f'(a)(x - a) + \frac{f''(a)}{2!}(x - a)^2 + \cdots + \frac{f^{(n)}(a)}{n!}(x - a)^n}_{\text{Taylor polynomial } P_n(x)}$$
$$+ \underbrace{\frac{f^{(n+1)}(c)}{(n + 1)!}(x - a)^{n+1}}_{\text{Remainder } R_n(x)}$$

* Some Famous Taylor Series:

$$e^x = 1 + x + \frac{x^2}{2!} + \frac{x^3}{3!} + \cdots$$

$$\sin x = x - \frac{x^3}{3!} + \frac{x^5}{5!} - \cdots$$

$$\cos x = 1 - \frac{x^2}{2!} + \frac{x^4}{4!} - \cdots$$

$$\frac{1}{1-x} = 1 + x + x^2 + \cdots \qquad \text{for any } |x| < 1$$

Vectors in the Plane:

* A vector that begins at (x_1, y_1) and ends at (x_2, y_2) is given by $\mathbf{v} = \langle x_2 - x_1, y_2 - y_1 \rangle$.

* If $\mathbf{v} = \langle x, y \rangle$, its *length* is given by $|\mathbf{v}| = \sqrt{x^2 + y^2}$.

* If $\mathbf{v} = \langle x_1, y_1 \rangle$ and $\mathbf{w} = \langle x_2, y_2 \rangle$, then

$$\mathbf{v} + \mathbf{w} = \langle x_1 + x_2, y_1 + y_2 \rangle$$

and

$$\mathbf{v} - \mathbf{w} = \langle x_1 - x_2, y_1 - y_2 \rangle$$

* $c\mathbf{a} = \langle ca_1, ca_2 \rangle$ is called *scalar multiplication*.

* Rules for Vectors:

$$\mathbf{a} + \mathbf{b} = \mathbf{b} + \mathbf{a}$$
$$\mathbf{a} + (\mathbf{b} + \mathbf{c}) = (\mathbf{a} + \mathbf{b}) + \mathbf{c}$$
$$c(\mathbf{a} + \mathbf{b}) = c\mathbf{a} + c\mathbf{b}$$
$$(c + e)\mathbf{a} = c\mathbf{a} + e\mathbf{a}$$
$$(ce)\mathbf{a} = c(e\mathbf{a}) = e(c\mathbf{a})$$

Vectors in Space:

* If $\mathbf{v} = \langle x, y, z \rangle$, then its length $|\mathbf{v}| = \sqrt{x^2 + y^2 + z^2}$.

* Given $\mathbf{v}$, $\mathbf{u} = \dfrac{\mathbf{v}}{|\mathbf{v}|}$ is a *unit vector* pointing in the same direction as $\mathbf{v}$.

* Three useful unit vectors are:

$$\mathbf{i} = \langle 1, 0, 0 \rangle$$
$$\mathbf{j} = \langle 0, 1, 0 \rangle$$
$$\mathbf{k} = \langle 0, 0, 1 \rangle$$

$$\mathbf{a} = \langle a_1, a_2, a_3 \rangle = a_1\mathbf{i} + a_2\mathbf{j} + a_3\mathbf{k}.$$

* The *dot product* of $a = \langle a_1, a_2, a_3 \rangle$ and $b = \langle b_1, b_2, b_3 \rangle$ is the number $\mathbf{a} \cdot \mathbf{b} = a_1b_1 + a_2b_2 + a_3b_3$.

$$\mathbf{a} \cdot \mathbf{a} = |\mathbf{a}|^2$$
$$\mathbf{a} \cdot \mathbf{b} = \mathbf{b} \cdot \mathbf{a}$$
$$c\mathbf{a} \cdot \mathbf{b} = \mathbf{a} \cdot c\mathbf{b} = c(\mathbf{a} \cdot \mathbf{b})$$
$$\mathbf{a} \cdot \mathbf{b} = |\mathbf{a}||\mathbf{b}|\cos\theta$$

* Vectors $\mathbf{a}$ and $\mathbf{b}$ are perpendicular if and only if $\mathbf{a} \cdot \mathbf{b} = 0$.

* The component (a length) of vector $\mathbf{a}$ along $\mathbf{b}$ is given by

$$\text{comp}_\mathbf{b}\, \mathbf{a} = |\mathbf{a}|\cos\theta$$
$$= |\mathbf{a}|\left(\frac{\mathbf{a} \cdot \mathbf{b}}{|\mathbf{a}||\mathbf{b}|}\right) = \frac{\mathbf{a} \cdot \mathbf{b}}{|\mathbf{b}|}$$

The projection (a vector) of vector $\mathbf{a}$ along $\mathbf{b}$ is given by

$$\text{proj}_\mathbf{b}\, \mathbf{a} = (\text{comp}_\mathbf{b}\, \mathbf{a})\frac{\mathbf{b}}{|\mathbf{b}|}$$
$$= \frac{\mathbf{a} \cdot \mathbf{b}}{|\mathbf{b}|^2}\mathbf{b}$$

Determinants:

$$\det\begin{pmatrix} a & b \\ c & d \end{pmatrix} = \begin{vmatrix} a & b \\ c & d \end{vmatrix} = ad - bc$$

$$\det\begin{pmatrix} a & b & c \\ d & e & f \\ g & h & j \end{pmatrix} = \begin{vmatrix} a & b & c \\ d & e & f \\ g & h & j \end{vmatrix} = a\begin{vmatrix} e & f \\ h & j \end{vmatrix} - b\begin{vmatrix} d & f \\ g & j \end{vmatrix} + c\begin{vmatrix} d & e \\ g & h \end{vmatrix}$$

Cross Product:

* The cross product of vectors $\mathbf{a}$ and $\mathbf{b}$ is given by

$$\mathbf{a} \times \mathbf{b} = \langle a_2 b_3 - a_3 b_2, a_3 b_1 - a_1 b_3, a_1 b_2 - a_2 b_1 \rangle = \begin{vmatrix} \mathbf{i} & \mathbf{j} & \mathbf{k} \\ a_1 & a_2 & a_3 \\ b_1 & b_2 & b_3 \end{vmatrix}$$

* $\mathbf{a} \times \mathbf{b}$ is perpendicular to $\mathbf{a}$ and $\mathbf{b}$.

* $\mathbf{a} \times \mathbf{b} = -\mathbf{b} \times \mathbf{a}$

* $|\mathbf{a} \times \mathbf{b}| = |\mathbf{a}||\mathbf{b}| \sin \theta$

* Area (parallelogram) $= |\mathbf{a}||\mathbf{b}| \sin \theta = |\mathbf{a} \times \mathbf{b}|$
  Area (triangle) $= \frac{1}{2} |\mathbf{a} \times \mathbf{b}|$

* $\mathbf{a} \cdot (\mathbf{b} \times \mathbf{c}) = \begin{vmatrix} a_1 & a_2 & a_3 \\ b_1 & b_2 & b_3 \\ c_1 & c_2 & c_3 \end{vmatrix}$

Lines in Space:

* Given one point $P(x_0, y_0, z_0)$ on a line and a vector $\mathbf{v} = \langle a, b, c \rangle$ in the direction of the line, the *parametric equations* for the line are

$$x = x_0 + ta$$
$$y = y_0 + tb$$
$$z = z_0 + tc$$

* *Symmetric equations* for the line:

$$\frac{x - x_0}{a} = \frac{y - y_0}{b} = \frac{z - z_0}{c}$$

Planes in Space:

Given a normal vector $\mathbf{n} = \langle a, b, c \rangle$ and a point $P_0 = (x_0, y_0, z_0)$ on the plane, the equation for the plane is

$$a(x - x_0) + b(y - y_0) + c(z - z_0) = 0$$

Parametric Curves:

* As t varies, $\mathbf{r}(t) = \langle x(t), y(t), z(t) \rangle$ gives a parametric curve in space. $\mathbf{r}(t)$ is called the *position vector*.

* The *velocity vector* $\mathbf{v} = \mathbf{r}'(t) = \langle x'(t), y'(t), z'(t) \rangle$ is tangent to the curve. $|\mathbf{v}|$ is the *speed*.

* The *acceleration vector* $\mathbf{a}(t) = \mathbf{v}'(t) = \mathbf{r}''(t)$.

* Rules for differentiating vector valued functions:

1. $\dfrac{d}{dt}[\mathbf{r}(t) + \mathbf{s}(t)] = \dfrac{d}{dt}\mathbf{r}(t) + \dfrac{d}{dt}\mathbf{s}(t)$

2. $\dfrac{d}{dt}[c\mathbf{r}(t)] = c\dfrac{d}{dt}\mathbf{r}(t)$

3. $\dfrac{d}{dt}[f(t)\mathbf{r}(t)] = f(t)\mathbf{r}'(t) + f'(t)\mathbf{r}(t)$

4. $\dfrac{d}{dt}[\mathbf{r}(t) \bullet \mathbf{s}(t)] = \mathbf{r}(t) \bullet \mathbf{s}'(t) + \mathbf{r}'(t) \bullet \mathbf{s}(t)$

5. $\dfrac{d}{dt}[\mathbf{r}(t) \times \mathbf{s}(t)] = \mathbf{r}(t) \times \mathbf{s}'(t) + \mathbf{r}'(t) \times \mathbf{s}(t)$

6. $\dfrac{d}{dt}[\mathbf{r}(f(t))] = \mathbf{r}'(f(t))f'(t)$

* To integrate a vector-valued function,

$$\int_a^b \mathbf{r}(t)\, dt = \left\langle \int_a^b x(t)\, dt, \int_a^b y(t)\, dt, \int_a^b z(t)\, dt \right\rangle$$

* Formula for the arc length of a curve:

$$S = \int_a^b |\mathbf{v}|\, dt = \int_a^b \sqrt{[x'(t)]^2 + [y'(t)]^2 + [z'(t)]^2}\, dt$$

Curvature:

* The curvature $\kappa(s)$ of a unit speed curve $\mathbf{r}(s)$ is $\kappa(s) = |\mathbf{r}''(s)|$.

* The curvature of an arbitrary curve $\mathbf{r}(t)$ is $\kappa(t) = \dfrac{|\mathbf{T}'(t)|}{|\mathbf{r}'(t)|}$, where $\mathbf{T}(t) = \dfrac{\mathbf{r}'(t)}{|\mathbf{r}'(t)|}$.

How to Ace the Rest of Calculus: Just the Facts

Surfaces:

* $ax + by + cz = d$ is a plane.

* $z = ax^2 + by^2$ is a paraboloid if a and b have the same sign.

* $\dfrac{x^2}{a^2} + \dfrac{y^2}{b^2} + \dfrac{z^2}{c^2} = 1$ is an ellipsoid with axis lengths a, b, and c.

* $z^2 = ax^2 + by^2 - c$ with $a, b, c > 0$ is a hyperboloid of one sheet opening on the z-axis.

* $z^2 = ax^2 + by^2 + c$ with $a, b, c > 0$ is a hyperboloid of two sheets opening on the z-axis.

* $z^2 = ax^2 + by^2$ with $a, b > 0$ is a pair of elliptic cones meeting at their vertices and opening on the z-axis.

* $z = ax^2 - by^2$ with $a, b > 0$ is a saddle (also called a hyperbolic paraboloid).

* We can switch x, y, and z to get surfaces opening along the other axes.

Limits and Continuity:

* $\displaystyle\lim_{(x,\,y)\to(a,\,b)} f(x, y) = L$ means that for every path in the xy-plane to $(a, b), f(x, y)$ approaches L.

* A function $f(x, y)$ is *continuous* at $(x, y) = (a, b)$ if $\displaystyle\lim_{(x,\,y)\to(a,\,b)} f(x, y) = f(a, b)$.

Partial Derivatives:

* The *partial derivative* of f with respect to x is $\dfrac{\partial f}{\partial x} = \displaystyle\lim_{h\to 0} \dfrac{f(x + h, y) - f(x, y)}{h}$.

* The *partial derivative* of f with respect to y is $\dfrac{\partial f}{\partial y} = \displaystyle\lim_{h\to 0} \dfrac{f(x, y + h) - f(x, y)}{h}$.

* $\dfrac{\partial f}{\partial x}(a, b)$ is the slope of the tangent line to the surface that lies in the plane that passes through the point $(a, b, f(a, b))$ and that is parallel to the xz-plane.

* $\dfrac{\partial f}{\partial y}(a, b)$ is the slope of the tangent line to the surface that lies in the plane that passes through the point $(a, b, f(a, b))$ and that is parallel to the yz-plane.

* To find $\dfrac{\partial f}{\partial x}$, treat y as a constant, and differentiate with respect to x.

  To find $\dfrac{\partial f}{\partial y}$, treat x as a constant, and differentiate with respect to y.

* When all these partial derivatives are continuous, $\dfrac{\partial^2 f}{\partial y \, \partial x} = \dfrac{\partial^2 f}{\partial x \, \partial y}$.

* At a local maximum or minimum, $\dfrac{\partial f}{\partial x} = 0$ and $\dfrac{\partial f}{\partial y} = 0$, if they exist.

Constrained Max-Min Problem:

To find the maximum or minimum of $f(x, y, z)$ when $g(x, y, z) = 0$:

Step 1 Set up the function $f(x, y, z)$ to be maximized and the so-called *constraint equation* $g(x, y, z) = 0$.

Step 2 Use the constraint equation to reduce the number of variables in f to two.

Step 3 Take the partial derivatives of f and set them equal to 0.

Step 4 Solve these equations for the critical points.

Step 5 Determine which if any of the critical points corresponds to a maximum or minimum.

Chain Rule:

If w depends directly on u and v, each of which depends on x and y, then

$$\frac{\partial w}{\partial x} = \frac{\partial w}{\partial u}\frac{\partial u}{\partial x} + \frac{\partial w}{\partial v}\frac{\partial v}{\partial x}$$

$$\frac{\partial w}{\partial y} = \frac{\partial w}{\partial u}\frac{\partial u}{\partial y} + \frac{\partial w}{\partial v}\frac{\partial v}{\partial y}$$

How to Ace the Rest of Calculus: Just the Facts

Directional Derivative:

* The *gradient* of a function $f(x, y)$ is given by $\nabla f = \left\langle \dfrac{\partial f}{\partial x}, \dfrac{\partial f}{\partial y} \right\rangle$.

* The *directional derivative* is $D_u f = \nabla f \cdot \mathbf{u}$.

* The direction of steepest ascent is in the direction of ∇f and $D_u f = |\nabla f|$ in this direction. The direction of steepest descent is in the direction of $-\nabla f$ and $D_u f = -|\nabla f|$ in this direction.

Lagrange Multipliers:

To find the maximum or minimum of $f(x, y, z)$ when $g(x, y, z) = 0$:

Step 1 Make sure the constraint is in the form $g(x, y, z) = 0$.

Step 2 Set $\nabla f = \lambda \nabla g$.

Step 3 Solve the resulting set of equations to find x, y, and z.

Step 4 Determine which of the solutions corresponds to the maximum or minimum.

Second Derivative Test:

The *discriminant* of f is given by $\Delta = f_{xx} f_{yy} - (f_{xy})^2$.
If (a, b) is a critical point of $f(x, y)$, then

1. If $\Delta > 0$ and $f_{xx} > 0$, then f has a local minimum at (a, b).

2. If $\Delta > 0$ and $f_{xx} < 0$, then f has a local maximum at (a, b).

3. If $\Delta < 0$, then f has neither a local max or min at (a, b) but rather has a saddle point there.

4. If $\Delta = 0$, then the second derivative test tells us absolutely nothing.

This test is encapsulated in these faces:

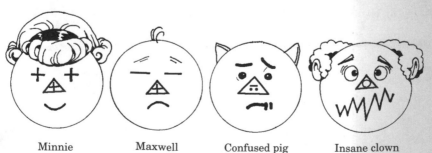

Minnie   Maxwell   Confused pig   Insane clown

Multiple Integrals:

* Double integrals over a region R are defined as limits of Riemann sums:

$$\iint_R f(x, y)\, dA = \lim_{n \to \infty} \sum_{i=1}^{n} f(x_i, y_i)\, \Delta A_i$$

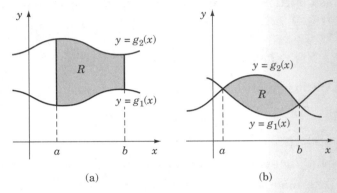

(a)   (b)

When R is the region under $y = g_2(x)$ and above $y = g_1(x)$ between $x = a$ and $x = b$, the double integral of $f(x, y)$ over R becomes

$$\int_a^b \left[\int_{g_1(x)}^{g_2(x)} f(x, y)\, dy \right] dx$$

Do the inner integral $\int_{g_1(x)}^{g_2(x)} f(x, y)\, dy$ first, treating x as a constant. Then take your answer, recast x as a variable, and do the outer integral.

When R is the region to the right of $x = h_1(y)$ and to the left of $x = h_2(y)$ between $y = c$ and $y = d$, the double integral of $f(x, y)$ over R becomes

$$\int_c^d \left[\int_{h_1(y)}^{h_2(y)} f(x, y)\, dx \right] dy$$

* Polar Coordinates: Points in the plane are described by (r, θ), and a double integral over a region R looks like $\iint_R f(r, \theta) r\, dr\, d\theta$.

* Triple integrals over a solid region V in 3-space are also defined in terms of a limit of Riemann sums:

$$\iiint_V f(x, y, z)\, dV = \lim_{n \to \infty} \sum_{i=1}^{n} f(x_i, y_i, z_i)\, \Delta V_i$$

A typical region V is described by setting $h_1(x, y) \leq z \leq h_2(x, y)$, $g_1(x) \leq y \leq g_2(x)$ and $a \leq x \leq b$. Then the integral takes the form

$$\int_a^b \left[\int_{g_1(x)}^{g_2(x)} \left[\int_{h_1(x, y)}^{h_2(x, y)} f(x, y, z)\, dz \right] dy \right] dx$$

It's calculated from the inside out, in a series of three integrations.

* Cylindrical Coordinates: Points in space are described by (r, θ, z), where $x = r \cos \theta$, $y = r \sin \theta$, $z = z$. Triple integrals over a solid region V take the form

$$\iiint_V f(r, \theta, z) r\, dr\, d\theta\, dz$$

* Spherical Coordinates: Points in space are described by (ρ, ϕ, θ), where $x = \rho \sin \phi \cos \theta$, $y = \rho \sin \phi \sin \theta$, $z = \rho \cos \phi$. Triple integrals over a solid region V take the form

$$\iiint_V f(\rho, \phi, \theta)\, \rho^2 \sin \phi\, d\rho\, d\phi\, d\theta$$

* Mass, Center of Mass, and Moments: The mass of a region V is obtained by integrating the density function σ over a region.

$$\text{Mass} = \iiint_V \sigma(x, y, z)\, dx\, dy\, dz$$

The *center of mass* of V, also called the *centroid*, is a point $(\bar{x}, \bar{y}, \bar{z})$ where a solid "balances." Its coordinates are given by

$$\bar{x} = \frac{\iiint_V \sigma(x, y, z)\, x\, dx\, dy\, dz}{\iiint_V \sigma(x, y, z)\, dx\, dy\, dz}$$

$$\bar{y} = \frac{\iiint_V \sigma(x, y, z)\, y\, dx\, dy\, dz}{\iiint_V \sigma(x, y, z)\, dx\, dy\, dz}$$

$$\bar{z} = \frac{\iiint_V \sigma(x, y, z)\, z\, dx\, dy\, dz}{\iiint_V \sigma(x, y, z)\, dx\, dy\, dz}$$

The integrals in the numerator are called the *first moments* of V.

* The *moment of inertia* of a solid V measures the resistance to getting it twirling. For twirling around the z-axis, it's given by

$$\iiint_V \sigma(x, y, z)(x^2 + y^2)\, dx\, dy\, dz$$

* Change of Coordinates: When we use some coordinates (u, v, w) other than the rectangular coordinates (x, y, z), we have to adapt our formula for the multiple integral over a region. If we express each of x, y, z as a function of (u, v, w), so $x = x(u, v, w), y = y(u, v, w), z = z(u, v, w)$, then let

$$\left| \frac{\partial(x, y, z)}{\partial(u, v, w)} \right| = \det \begin{bmatrix} \dfrac{\partial x}{\partial u} & \dfrac{\partial y}{\partial u} & \dfrac{\partial z}{\partial u} \\[2mm] \dfrac{\partial x}{\partial v} & \dfrac{\partial y}{\partial v} & \dfrac{\partial z}{\partial v} \\[2mm] \dfrac{\partial x}{\partial w} & \dfrac{\partial y}{\partial w} & \dfrac{\partial z}{\partial w} \end{bmatrix}$$

The triple integral in u-, v-, w-coordinates over a region V is

$$\iiint_V f(u, v, w) \left| \frac{\partial(x, y, z)}{\partial(u, v, w)} \right| du\, dv\, dw$$

Vector Fields, Div, and Curl:

* A vector field in the plane gives a vector at each point in the plane, and looks like $\mathbf{W}(x, y) = P(x, y)\mathbf{i} + Q(x, y)\mathbf{j}$. A vector field in space looks like like $\mathbf{F}(x, y, z) = P(x, y, z)\mathbf{i} + Q(x, y, z)\mathbf{j} + R(x, y, z)\mathbf{k}$.

* A vector field $\mathbf{F}$ which is the gradient of a function f is called *conservative*, and f is called a *potential* function.

* The *divergence* of a vector field $\mathbf{F}$ is a function that is given by the formula $\operatorname{div} \mathbf{F} = \dfrac{\partial P}{\partial x} + \dfrac{\partial Q}{\partial y} + \dfrac{\partial R}{\partial z}$. The divergence is positive at points where $\mathbf{F}$ is expanding and negative at points where $\mathbf{F}$ is compressing. A shorthand form for its equation is $\operatorname{div} \mathbf{F} = \nabla \cdot \mathbf{F}$, where ∇ is the operator

$$\nabla = \frac{\partial}{\partial x}\mathbf{i} + \frac{\partial}{\partial y}\mathbf{j} + \frac{\partial}{\partial z}\mathbf{k}$$

* The *curl* of a vector field measures the tendency of a vector field to twirl things around an axis. For a vector field $\mathbf{F}(x, y, x) = P(x, y, z)\mathbf{i} + Q(x, y, z)\mathbf{j} + R(x, y, z)\mathbf{k}$, we get curl $\mathbf{F}$ by taking the cross product of ∇ and $\mathbf{F}$:

$$\text{curl } \mathbf{F} = \nabla \times \mathbf{F} = \det \begin{bmatrix} \frac{\partial}{\partial x} & \frac{\partial}{\partial y} & \frac{\partial}{\partial z} \\ P & Q & R \end{bmatrix}$$

* Line Integrals: We use these to integrate along a curve. When a curve C is described by the parametrization $\mathbf{r}(t) = \langle x(t), y(t), z(t) \rangle$, $a \le t \le b$, the arc length of the curve is given by

$$\int_C ds = \int_C |\mathbf{v}| \, dt = \int_a^b \sqrt{[x'(t)]^2 + [y'(t)]^2 + [z'(t)]^2} \, dt$$

To calculate work done against a force, we calculate a line integral using the force field $\mathbf{F}$:

$$\text{Work} = \int_C \mathbf{F} \cdot d\mathbf{r} = \int_a^b \mathbf{F}(\mathbf{r}(t)) \cdot \mathbf{r}'(t) \, dt$$

* Properties of Conservative Vector Fields:

  1. The line integral of a conservative vector field along a curve can be calculated by taking the difference of its potential function at the two endpoints of the curve.

  2. The line integral of a vector field which is conservative in a region depends only on the the curve's starting and ending points. It is *independent of path*.

  3. The line integral of a conservative vector field in a region around a loop in that region equals zero.

  4. A conservative vector field always satisfies $\frac{\partial P}{\partial y} = \frac{\partial Q}{\partial x}$.

* If $\frac{\partial P}{\partial y} = \frac{\partial Q}{\partial x}$ on a *simply connected* region R, then $F = \nabla f$ for some function $f(x, y)$ on R, and so F is conservative on R.

* Green's Theorem: If $F = P\mathbf{i} + Q\mathbf{j}$ is a vector field and C goes around a region D, then

$$\oint_C P \, dx + Q \, dy = \iint_D \left(\frac{\partial Q}{\partial x} - \frac{\partial P}{\partial y} \right) dx \, dy$$

The *flux* of a vector field across a curve measures how much of the vector field is flowing across the curve. Calculate a unit length normal vector **n** to a curve. For a curve $\mathbf{r}(t) = \langle x(t), y(t) \rangle$,

$$\mathbf{n} = \frac{y'(t)}{|\mathbf{r}'(t)|}\mathbf{i} + \frac{-x'(t)}{|\mathbf{r}'(t)|}\mathbf{j}$$

The flux is given by

$$\text{Flux} = \oint_C \mathbf{F} \cdot \mathbf{n}\, ds$$

* The Divergence Theorem: The divergence of a vector field in a region R with boundary curve C in the plane equals the total flow of the vector field across the boundary, or the flux.

$$\iint_R \text{div } \mathbf{F}\, dx\, dy = \oint_C \mathbf{F} \cdot \mathbf{n}\, ds$$

* Surface Integrals: The surface area of a surface S in space given by $\mathbf{r}(u, v)$ is

$$\text{Area}(S) = \iint_S dS = \iint_S |\mathbf{r}_u \times \mathbf{r}_v|\, du\, dv$$

To integrate a function σ over a surface S in space, we take

$$\iint_S \sigma(x(u, v), y(u, v), z(u, v)) |\mathbf{r}_u \times \mathbf{r}_v|\, du\, dv$$

* Divergence Theorem in Space: The total divergence of a vector field in a region V in 3-space equals the flow across the boundary surface S.

$$\iiint_V \text{div } \mathbf{F}\, dx\, dy\, dz = \iint_S \mathbf{F} \cdot \mathbf{n}\, dS$$

* Stokes's Theorem: Given a surface S in space with unit normal vector **n** and boundary curve C, and a vector field **F** on S, the theorem equates a line integral over C and a surface integral over S.

$$\oint_C \mathbf{F} \cdot d\mathbf{r} = \iint_S \text{curl } \mathbf{F} \cdot \mathbf{n}\, dS$$